Springer Theses

Recognizing Outstanding Ph.D. Research

Aims and Scope

The series "Springer Theses" brings together a selection of the very best Ph.D. theses from around the world and across the physical sciences. Nominated and endorsed by two recognized specialists, each published volume has been selected for its scientific excellence and the high impact of its contents for the pertinent field of research. For greater accessibility to non-specialists, the published versions include an extended introduction, as well as a foreword by the student's supervisor explaining the special relevance of the work for the field. As a whole, the series will provide a valuable resource both for newcomers to the research fields described, and for other scientists seeking detailed background information on special questions. Finally, it provides an accredited documentation of the valuable contributions made by today's younger generation of scientists.

Theses may be nominated for publication in this series by heads of department at internationally leading universities or institutes and should fulfill all of the following criteria

- They must be written in good English.
- The topic should fall within the confines of Chemistry, Physics, Earth Sciences, Engineering and related interdisciplinary fields such as Materials, Nanoscience, Chemical Engineering, Complex Systems and Biophysics.
- The work reported in the thesis must represent a significant scientific advance.
- If the thesis includes previously published material, permission to reproduce this must be gained from the respective copyright holder (a maximum 30% of the thesis should be a verbatim reproduction from the author's previous publications).
- They must have been examined and passed during the 12 months prior to nomination.
- Each thesis should include a foreword by the supervisor outlining the significance of its content.
- The theses should have a clearly defined structure including an introduction accessible to new PhD students and scientists not expert in the relevant field.

Indexed by zbMATH.

Ralf Kundel

Accelerating Network Functions Using Reconfigurable Hardware

Design and Validation of High-Throughput and Low-Latency Network Functions at the Access Edge

Doctoral Thesis accepted by
Technical University of Darmstadt, Darmstadt, Germany

Springer

Author
Dr.-Ing. Ralf Kundel
Multimedia Communications Lab
TU Darmstadt
Darmstadt, Germany

Supervisors
Prof. Dr.-Ing. Dr. h.c. Ralf Steinmetz
Multimedia Communications Lab
TU Darmstadt
Darmstadt, Germany

Prof. Dr.-Ing. Carsten Griwodz
Department of Informatics
University of Oslo
Oslo, Norway

Prof. Dr. Boris Koldehofe
Distributed and Operating Systems Group
Technische Universität Ilmenau
Ilmenau, Germany

ISSN 2190-5053 ISSN 2190-5061 (electronic)
Springer Theses
ISBN 978-3-031-52871-2 ISBN 978-3-031-52872-9 (eBook)
https://doi.org/10.1007/978-3-031-52872-9

This Springer imprint is published by the registered company Springer Nature Switzerland AG
The registered company address is: Gewerbestrasse 11, 6330 Cham, Switzerland

Supervisors' Foreword

It is a great pleasure to introduce Dr. Ralf Kundel's thesis work, accepted for publication within Springer Theses and awarded with a prize for the best dissertation of the Department of Electrical Engineering and Information Technology by the association of friends of the Technical University of Darmstadt, known as Vereinigung von Freunden der TU Darmstadt.

Dr. Kundel joined the Multimedia Communications Lab (KOM) at TU Darmstadt in January 2018 under the guidance of us, Ralf Steinmetz and Boris Koldehofe, with mentoring from Carsten Griwodz. Dr. Kundel conducted his doctoral research, which provided essential components to the *Dynamic Networks* Project in collaboration with the Deutsche Telekom. He was also affiliated with the Collaborative Research Center MAKI, funded by the Deutsche Forschungsgemeinschaft (DFG).

From the beginning, his enthusiasm, excellence in research, and determination were outstanding. His enthusiasm for programmable hardware, especially FPGAs and P4-programmable switches, laid the foundation early on for his further scientific career, which Dr. Kundel pursues with enormous speed and outstanding scientific precision.

Within Dr. Kundel's thesis, we want to emphasize the following notable contributions:

Initially, with the introduction of the *host bypassing* concept, he proposed a novel concept of hardware accelerator integration for network functions in commodity servers. While keeping the flexibility high, host bypassing substantially lowers the latency variation and host CPU utilization. This approach enables and supports the execution of network functions with high Quality of Service (QoS) needs, which we encounter nowadays in 5G networks.

Secondly, he introduced techniques for mapping network functions to create fixed and mobile Internet access in programmable hardware accelerators encompassing 5G networks. The remarkable outcomes demonstrate latency improvements of nearly two magnitudes while improving the throughput of those network functions by a factor of five compared to state-of-the-art approaches.

Thirdly, in order to enhance the end-to-end quality of service in access networks, he demonstrated the feasibility of Active Queue Management (AQM) mechanisms in conjunction with hardware-accelerated network functions. This strategy has the potential to improve the network latency of the future Internet substantially.

Lastly, he observed the challenge of high-accuracy evaluations of network functions in his work, especially in terms of time accuracy, loss detection, and the simultaneous generation of high loads. In response, he introduced the *P4STA* framework, utilizing his innovative concept of disaggregated load generation, timestamping, and evaluation. Through this approach, the latency of network packets at line rate, presently reaching up to 100 Gbit/s, can be measured with an accuracy of approximately 2 ns.

In summary, the contributions and results of Dr. Kundel's outstanding thesis have demonstrated the significant potential of hardware acceleration in forthcoming access edge network functions, as well as the methodologies for their integration with Internet access networks. We hope the interested reader will enjoy the excellent motivation and solutions to core problems in today's residential and mobile access to the Internet.

Darmstadt, Germany — Prof. Dr.-Ing. Dr. h.c. Ralf Steinmetz
Illmenau, Germany — Prof. Dr. Boris Koldehofe
Oslo, Norway — Prof. Dr.-Ing. Carsten Griwodz
September 2023

Previously Published Material

This thesis includes material previously published in scientific journals, magazines, conference proceedings, and patent applications. Even though this thesis does not contain any word-for-word copy of the previously published material, contents are reused. This includes adopted or reprocessed scientific concepts, figures, tables, and evaluation results.

Parts of this thesis have been published in the following documents:

- Ralf Kundel et al. “Chaining of Hardware Accelerated Virtual Network Functions in PCIe Environments.” In: *Proceedings of the 20th International Middleware Conference Demos and Posters.* Middleware ’19. ACM, Dec. 2019, pp. 13–14
- Ralf Kundel et al. “Host Bypassing: Direct Data Piping from the Network to the Hardware Accelerator.” In: *Proceedings of the 14th International Symposium on Embedded Multicore/Many-core Systems-on-Chip (MCSoC).* Best Paper Award. IEEE, Dec. 2021, pp. 23–30
- Ralf Kundel et al. “Host Bypassing: Let your GPU speak Ethernet.” In: *Proceedings of the 8th International Conference on Network Softwarization (NetSoft).* Workshop on Edge Network Softwarization (ENS). IEEE, June 2022, pp. 1–6
- Ralf Kundel et al. “P4-BNG: Central Office Network Functions on Programmable Packet Pipelines.” In: *Proceedings of the 15th International Conference on Network and Service Management (CNSM).* IEEE, Oct. 2019, pp. 1–9
- Ralf Kundel et al. “OpenBNG: Central office network functions on programmable data plane hardware.” In: *International Journal of Network Management* 31.1 (Jan. 2021), pp. 1–25

- Ralf Kundel et al. "User Plane Hardware Acceleration in Access Networks: Experiences in Offloading Network Functions in Real 5G Deployments." In: *Proceedings of the 55th Hawaii International Conference on System Sciences.* ScholarSpace, Jan. 2022, pp. 1–10
- Ralf Kundel et al. "P4-CoDel: Active Queue Management in Programmable Data Planes." In: *Proceedings of the IEEE Conference on Network Function Virtualization and Software Defined Networks (NFV-SDN).* IEEE, Nov. 2018, pp. 1–4
- Ralf Kundel et al. "P4-CoDel: Experiences on Programmable Data Plane Hardware." In: *Proceedings of the IEEE International Conference on Communications (ICC).* IEEE, June 2021, pp. 1–6
- Ralf Kundel et al. "FPGA-assisted Massive Packet Queueing and Traffic Shaping at the Network Edge." In: *Proceedings of the 30th Annual International Symposium on Field-Programmable Custom Computing Machines (FCCM).* IEEE, May 2022, p. 1
- Ralf Kundel et al. "P4STA: High Performance Packet Timestamping with Programmable Packet Processors." In: *Proceedings of the IEEE/IFIP Network Operations and Management Symposium (NOMS).* IEEE, Apr. 2020, pp. 1–9
- Ralf Kundel et al. "Network Testing Utilizing Programmable Network Hardware." In: *IEEE Communications Magazine* (Feb. 2022), pp. 12–17
- Ralf Kundel, Amr Rizk, and Boris Koldehofe. "Microbursts in Software and Hardware-based Traffic Load Generation." In: *Proceedings of the IEEE/IFIP Network Operations and Management Symposium (NOMS).* IEEE, Apr. 2020, pp. 1–2
- Ralf Kundel, Fridolin Siegmund, and Boris Koldehofe. "How to measure the speed of light with programmable data plane hardware?" In: *Proceedings of the ACM/IEEE Symposium on Architectures for Networking and Communications Systems (ANCS).* IEEE, Sept. 2019, pp. 1–2
- Ralf Kundel et al. "Queueing at the Telco Service Edge: Requirements, Challenges and Opportunities." In: *Proceedings of the 1st Workshop on Buffer Sizing.* Stanford University, Dec. 2019, pp. 1–6

Patent Applications

- Ralf Kundel, Leonhard Nobach, and Fabian Schneider. *Method for an improved traffic shaping and/or management of ip traffic in a packet processing system, telecommunications network, system, program and computer program product.*

European Patent Office Application Number: 19215704.8-1215. Deutsche Telekom AG. Dec. 2019

- Ralf Kundel and Leonhard Nobach. *Method for an improved traffic shaping and/or management of ip traffic in a packet processing system, telecommunications network, network node or network element, program and computer program product*. European Patent Office Application Number: 20157453.0-1215. Deutsche Telekom AG. Feb. 2020
- Ralf Kundel and Leonhard Nobach. *Method for an improved traffic shaping and/or management of ip traffic in a packet processing system, telecommunications network, network node or network element, program and computer program product*. European Patent Office Application Number: 20157452.2-1215. Deutsche Telekom AG. Feb. 2020
- Ralf Kundel and Andreas Zimber. *Verfahren zum Erfassen und Verarbeiten eines Zahlsignals*. European Patent Office Application Number: 20213595.0—1215. Deutsche Telekom AG. Dec. 2020

Contents

Acronyms

3GPP	3rd Generation Partnership Project
AMF	Access And Mobility Management Function
AN	Access Node
API	Application Programming Interface
AQM	Active Queue Management
ASIC	Application Specific Integrated Circuit
BNG	Broadband Network Gateway
BRAM	Block Random Access Memory
CLI	Command Line Interface
CoDel	Controlled Delay
CPU	Central Processing Unit
DMA	Direct Memory Access
DPDK	Data Plane Development Kit
DRAM	Dynamic Random Access Memory
DUT	Device Under Test
FIFO	First In First Out
FPGA	Field Programmable Gate Array
FSM	Finite-State Machine
FTTB	Fiber To The Building
FTTC	Fiber To The Curb
FTTH	Fiber To The Home
GPU	Graphics Processing Unit
gRPC	Google Remote Procedure Call
GTP	GPRS Tunneling Protocol
HQoS	Hierarchical Quality Of Service
IP core	Intellectual Property Core
ISP	Internet Service Provider
LUT	Lookup Table
MAC	Media Access Controller
MPLS	Multiprotocol Label Switching
MSAN	Multi-Service Access Node

MTU	Maximum Transmission Unit
NAT	Network Address Translation
NFV	Network Functions Virtualization
NIC	Network Interface Card
NPU	Network Processing Unit
OLT	Optical Line Terminal
P4	Programming Protocol-Independent Packet Processors
PCIe	Peripheral Component Interconnect Express
PPPoE	Point-to-Point Protocol Over Ethernet
QoS	Quality of Service
RAN	Radio Access Network
RED	Random Early Detection
RG	Residential Gateway
RTT	Round-Trip Time
SBA	Service-Based Architecture
SDN	Software Defined Networking
SmartNIC	Smart Network Interface Card
SMF	Session Management Function
SRAM	Static Random Access Memory
TCP	Transmission Control Protocol
TEID	Tunnel Endpoint ID
UE	User Equipment
UPF	User Plane Function
vDSL	Very High-Speed Digital Subscriber Line
VLAN	Virtual LAN
VoIP	Voice Over IP

Chapter 1
Introduction

During the last decades, digital communication networks have become the basis of many applications in everyday life. This digital transformation process will most likely continue in the future, as supposedly additional services will be built upon digital computer networks, such as autonomous driving, smart manufacturing, or medical care [2, 8, 33]. The challenges during the COVID pandemic and the accompanying rise of remote work have shown the importance of reliable and performant communication networks. Connecting billions of people worldwide via digital video presence requires low latency and high throughput connectivity.

But even before the COVID-19 pandemic, the Internet has developed into a gigantic network that continues to grow. In 2021, more than five billion people, around two-thirds of the whole mankind, are connected to this huge network with even more devices. Besides the number of users, the bandwidth consumption has increased tremendously. Between 1983 and 2019, the speed of Internet access lines, which means vDSL connections or equivalents, has a constant growth rate of 50% per year [30]. The increasing available bandwidth and decreasing latency resulted in many new applications upon the Internet in the past. For example, 360° video streaming or multi-player computer games became possible [5, 35].

Due to the increasing application requirements, providing the underlying network infrastructure is still challenging. Besides the steadily increasing network throughput, additional metrics obtained importance in the last years. For example, high availability and low latency are required in industrial automation networks [10]. Moreover, many end-user applications require not only high bandwidth but also low latency simultaneously [31]. To fulfill the novel demands of current and future applications, computer networks must provide high functional flexibility while having a satisfactory performance, characterized in general by a good Quality of Service (QoS). In this context, flexibility includes the ability to launch novel functionality, such as protocols or mechanisms, by reconfiguring the existing network hardware.

R. Kundel, *Accelerating Network Functions Using Reconfigurable Hardware*,
Springer Theses, https://doi.org/10.1007/978-3-031-52872-9_1

Furthermore, this large-scale connectivity in the Internet consumes a lot of energy. According to several studies, e.g., the article presented by Jones in 2018 [9], the worldwide energy consumption of the Internet is about 10% of the total electrical energy consumption and will rise further in the future. However, the author mentioned that by novel technologies, such as new processor architectures, the energy consumption of data centers can be reduced. In the same way, other technological advances can also assist in reducing the energy consumption of computer networks.

Consequently, it is of tremendous importance to achieve the required functional flexibility and performance in an energy-efficient manner. There is a huge necessity for research on the underlying hardware and corresponding utilization concepts, which can strongly contribute to achieving the aforementioned performance goals.

1.1 Motivation for Network Functions Acceleration

The constantly growing and varying requirements on computer networks require high flexibility and increasing performance. Software Defined Networking (SDN) and Network Functions Virtualization (NFV) propose a toolset of concepts providing increased programmability and execution flexibility to address this demand. The SDN concept introduces an improved configuration interface for network devices by exposing an open configuration API [7, 29]. SDN improves the flexibility in computer networks by disaggregating the hardware switches, defined as the *data plane*, from the network *control plane*. In extension to this, NFV enables the programming and the flexible execution of network functions by deploying the required functionality as a software component in universal compute nodes. These compute nodes are off-the-shelf data center servers and have good network connectivity [4].

While conventional networking hardware and its network functions are built upon Application Specific Integrated Circuits (ASICs), providing very high performance and good energy efficiency, they suffer from limited flexibility and slow innovation speed [6].

In contrast, network functionality built upon the NFV approach, running on commodity off-the-shelf hardware, has significantly higher flexibility; however, the achievable performance per device is a magnitude lower than for conventional ASIC-based solutions [32, 34]. This results in a large number of stacked NFV-servers providing the same amount of network packet processing as a single conventional ASIC-based networking device. Further, this approach implies a strongly lowered throughput while having the same number of network devices.

Especially in Internet Service Provider (ISP) access networks, high flexibility and performance are required at the same time to fulfill the needs of millions of customers. Figure 1.1 depicts the main service creation scenarios. Most residential subscribers are connected via a Very High-Speed Digital Subscriber Line (vDSL) connection and a Multi-Service Access Node (MSAN) as an intermediate agent to the BNG. The Broadband Network Gateway (BNG) is the termination node for thousands of parallel subscriber connections. Analogous to the BNG, the User Plane Function (UPF)

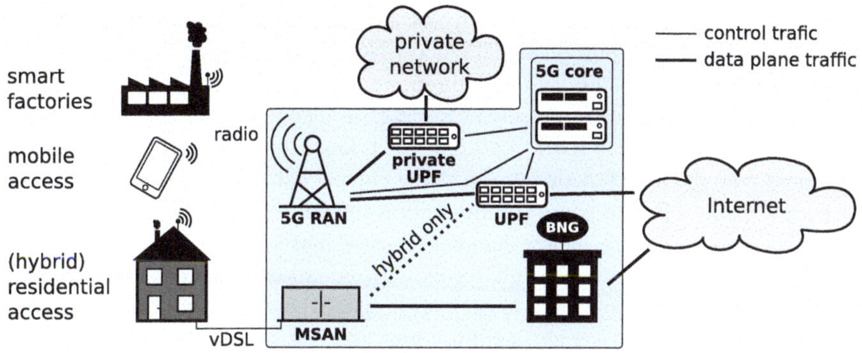

Fig. 1.1 Service creation for (1) residential access, (2) mobile access, and (3) industrial campus networks upon softwarized network infrastructure

is responsible for terminating all subscriber sessions in mobile access networks. Both network functions, the BNG and UPF, process each packet transmitted to or from the subscriber. Further, they suffer under the tradeoff between flexibility and performance. Assuming novel access technologies such as hybrid residential access, providing dual connectivity via vDSL and radio simultaneously, or network slicing for private networks, the demand for high performant and flexible network functions is even higher. Similarly, network functions in Radio Access Networks (RANs) have very high performance and latency requirements for en-/decoding each packet sent over the radio interface, especially in disaggregated OpenRAN environments.

To overcome the limitations of either inflexible ASIC-based solutions or functionality provided by less efficient NFV software appliances, programmable hardware constitutes a reasonable compromise, combining high flexibility and good performance. For example, Field Programmable Gate Arrays (FPGAs), Graphics Processing Units (GPUs), or P4-programmable ASICs can be utilized [3].

In this work, we investigate prospects of programmable hardware to improve the flexibility and performance of network functions in Internet access networks simultaneously.

1.2 Research Challenges

The aforementioned motivation imposes several challenges on the design, implementation, and validation of network functions. As part of this work, we identify and address two key challenges in the domain of ISP access networks:

Challenge: *Network functions at the Internet access edge must provide high flexibility and functionality while providing high performance and low innovation cycle times.* Implementing network functions as a fixed-function ASIC provides, as known today, the best achievable performance. However, these fixed-function chips have the big drawback of slow innovation cycles and immense development costs. For example, the previously introduced UPF and BNG in Internet access networks must provide specific functionality and high throughput at the same time. In this context, introducing new protocols and mechanisms, e.g., as a new functionality of the latest mobile radio standard *5G (3GPP release 15)* [1], leads Internet service providers to either integrate the new functionality into the current hardware or renew the entire hardware. While software-based solutions allow updating the current hardware flexibly, they cannot satisfy the heavy demand for good performance. Therefore, a huge need for flexibly programmable platforms with good performance exists. Consequently, novel concepts to benefit from programmable hardware must be established to simultaneously achieve high flexibility and good deterministic performance in Internet access network functions.

Challenge: *The validation of high-performance network functions is vital, yet precisely evaluating their functional and performance characteristics under load is challenging.* As addressed by the first research challenge, the demand for high-performance network functions requires QoS performance identifiers. However, the validation and verification of high-performance network functions demands measurement equipment being able to provide at least the same performance as the network function under test. As the performance level of the newest networking hardware is at the technological limit of chip manufacturing technology, meeting this performance by the test equipment is challenging. On the one hand, building special-purpose testing hardware would cause an immense effort and is not economical. On the other hand, inferior measurement equipment would not allow a high testing accuracy, i.e., a nanosecond time accuracy at 100 $Gbit/s$ throughput at potentially small packet sizes down to 64 *bytes*. In summary, no solution that provides high flexibility and accuracy for benchmarking network functions with a high performance exists.

1.3 Research Goals and Contributions

In the following, we present two research goals derived from the challenges above. As an implication of the first research challenge, the first research goal addresses how to put high-performance network functions into practice. Analogous, the second goal addresses the second challenge and focuses on validating such high-performance network functions. Each goal is made more specific by research questions answered within this work.

Research Goal 1: *Acceleration of network functions to meet the desired high-performance requirements of Internet access networks while providing flexibility by adaptability.*

This goal can be achieved by addressing the following two research questions:

RQ1.1: *How to design, implement, and integrate hardware accelerators, fulfilling the postulated QoS requirements, in networking edge environments?*

The integration into the data path is an integral part of hardware-accelerated network functions. In access networks, this path forwards all traffic to and from the subscriber, wherefore, a high performance, including high throughput and constant low latency, is required. However, some accelerator technologies, such as GPUs and FPGAs, do not offer direct network connectivity and must be integrated via the system bus of a host system, causing additional overhead. In addition to the integration, the design and implementation of a network function within a hardware accelerator can strongly affect its performance and is therefore of tremendous importance.

Contributions: To address this research question, we first introduce and investigate the *host bypassing* concept for a better data plane integration of FPGAs and GPUs, not providing any direct network interface, in 5G-RAN functions [11, 13, 14]. Further, we focus on the implementation of residential and mobile Internet subscriber termination in this context [15, 18]. In this context, we investigate the capability of integrating FPGAs as a QoS co-processor to programmable switches in Internet access networks [19].

Last, we focus on the expressiveness and capabilities of the P4 language to realize Active Queue Management (AQM), on the example of the CoDel algorithm [12, 22].

RQ1.2: *Which hardware constraints must be considered for selecting programmable off-the-shelf hardware technologies and algorithms for offloading network functions?*

Utilizing programmable hardware for offloading network functions implies several challenges. Due to multiple hardware constraints, the straightforward transfer of existing algorithms on hardware from a software reference implementation is not possible. For example, current programmable networking switches allow only non-iterative algorithms and basic arithmetic operations. In contrast to programmable switches, FPGAs allow iterative algorithms but offer a lower maximum bandwidth. Therefore, it is necessary to evaluate different programmable hardware platforms regarding their restrictions and their matching with the QoS requirements of the desired network function. Due to the different characteristics of different network functions, this consideration must be performed again for each application use case. Different network applications may benefit from other hardware platforms; thus, no single platform fits best for all applications.

Contributions: To address this research question, we investigate multiple programmable hardware platforms in this work for their application-specific eligibility. For that, one focus is on the expressiveness and capabilities of the domain-specific language P4 and corresponding hardware. In total, we investigate Graphics Processing Units (GPUs), Field Programmable Gate Arrays (FPGAs), and three different P4-programmable hardware platforms for various applications and scenarios of Internet service creation.

Specifically, we investigate the capabilities and performance characteristics of these three P4 platforms and FPGAs for the termination of Internet subscribers [15, 18–20]. Further, we investigate the functional and performance differences between GPUs and FPGAs integrated into the data plane of mobile access networks [11, 14].

Research Goal 2: *Validation of high-performance network functions.*

We address the following research question in this thesis to achieve the second research goal:

RQ2: *How to accomplish the high measurement accuracy requirements with programmable hardware for flexible network function testing?*

High-accuracy measurements are crucial to validate the correct behavior of network functions with the highest performance. While hardware-based measurement approaches suffer under limited flexibility, software approaches do not provide the required accuracy. Analogous to *RQ1.2*, the choice of measurement hardware determines the maximum possible time accuracy and throughput.

Contributions: To address this research question, we investigate the conjunction of both concepts, software- and hardware-based network function benchmarking, to benefit from both. Based on our presented framework, named *P4STA*, we investigate the advantages of a disaggregated testing environment [24, 25]. In this work, we investigate two hardware platforms and varying internal configurations, such as the best location for retrieving timestamps and performing loss detection, to determine the maximum achievable performance [23, 26].

Scope Differentiation

In this thesis, we focus on the acceleration of network functions. The approaches presented in this thesis will aim primarily at improving the flexibility and performance at the Internet access edge. At the same time, these approaches commonly enhance the energy efficiency compared to state-of-the-art methods. Nevertheless, this is only a side effect of our work and, thus, not analyzed in detail. Further, the introduction of novel concepts for describing and implementing network functions imposes many new challenges for reliability and security, which are also not part of this thesis. Last, we do not focus on the management and orchestration of hardware accelerators.

1.4 Structure of the Thesis

The outline of this thesis is as follows: First, in Chap. 2, the background and related work are discussed. In Chap. 3, we present the contributions belonging to the first research goal, acceleration network functions to fulfill the required QoS performance identifiers. To validate these contributions, we introduce the novel measurement methodology and framework *P4STA* in Chap. 4, addressing the second research goal. In the same chapter, we validate the correctness of this introduced measurement framework.

In the evaluation of this thesis, presented in Chap. 5, we evaluate the contributions of network function acceleration in the subsequent evaluation sections. Finally, in Chap. 6, the thesis is concluded with a summary of the main contributions, and we provide an outlook on potential future work.

References

1. 3GPP. Release 15 Description; Summary of Rel-15 Work Items. Technical report (TR). Version 15.0.0. Oct. 2019
2. Alemdar H, Ersoy C (2010) Wireless sensor networks for healthcare: a survey. Comput. Netw. 54(15):2688–2710
3. Bosshart P, Daly D, Gibb G, Izzard M, McKeown N, Rexford J, Schlesinger C, Talayco D, Vahdat A, Varghese G, Walker D (2014) P4: programming protocol-independent packet processors. ACM SIGCOMM Comput Commun Rev 44(3):87–95
4. Chiosi M, Clarke D, Willis P, Reid A et al. (2012) Network functions virtualisation: an introduction, benefits, enablers, challenges and call for actions. In: SDN and OpenFlow world congress, pp 1–16
5. Dasari M, Bhattacharya A, Vargas S, Sahu P, Balasubramanian A, Das SR (2020) Streaming 360-degree videos using super-resolution. In: IEEE INFOCOM 2020-IEEE conference on computer communications. IEEE, pp 1977–1986
6. Fu S, Liu J, Zhu W (2017) Multimedia content delivery with network function virtualization: the energy perspective. In: IEEE MultiMedia 24(3):38–47
7. Haleplidis E, Pentikousis K, Denazis S, Hadi J Salim, Meyer D, Koufopavlou O (2015) Software-defined networking (SDN): layers and architecture terminology. RFC 7426. Jan 2015. https://www.rfc-editor.org/info/rfc7426
8. Hobert L, Festag A, Llatser I, Altomare L, Visintainer F, Kovacs A (2015) Enhancements of V2X communication in support of cooperative autonomous driving. IEEE Commun Mag 53(12):64–70
9. Jones N (2018) How to stop data centres from gobbling up the world's electricity. Nature 561(7722):163–167
10. Khoshnevisan M, Joseph V, Gupta P, Meshkati F, Prakash R, Tinnakornsrisuphap P (2019) 5G industrial networks with CoMP for URLLC and time sensitive network architecture. IEEE J Select Areas Commun 37(4):947–959
11. Kundel R, Anderweit L, Markussen J, Griwodz C, Abboud O, Becker B, Meuser T (2022) Host bypassing: let your GPU speak ethernet. In: Proceedings of the 8th international conference on network softwarization (NetSoft). Workshop on edge network softwarization (ENS). IEEE, pp 1–6
12. Kundel R, Blendin J, Viernickel T, Koldehofe B, Steinmetz R (2018) P4-CoDel: active queue management in programmable data planes. In: Proceedings of the IEEE conference on network function virtualization and software defined networks (NFV-SDN). IEEE, pp 1–4
13. Kundel R, Burkert T, Griwodz C, Koldehofe B (2019) Chaining of hardware accelerated virtual network functions in PCIe environments. In: Proceedings of the 20th international middleware conference demos and posters. Middleware '19. ACM, pp 13–14
14. Kundel R, Eryigit K, Markussen J, Griwodz C, Abboud O, Hark R, Steinmetz R (2021) Host bypassing: direct data piping from the network to the hardware accelerator. In: Proceedings of the 14th international symposium on embedded Multicore/Many-core Systems-on-Chip (MCSoC). Best Paper Award. IEEE, pp 23–30
15. Kundel R, Meuser T, Koppe T, Hark R, Steinmetz R (2022) User plane hardware acceleration in access networks: experiences in offloading network functions in real 5G deployments. In: Proceedings of the 55th Hawaii international conference on system sciences. ScholarSpace, pp 1–10
16. Kundel R, Nobach L (2020) Method for an improved traffic shaping and/or management of ip traffic in a packet processing system, telecommunications network, network node or network element, program and computer program product. European patent office application number: 20157453.0-1215. Deutsche Telekom AG. Feb 2020
17. Kundel R, Nobach L (2020) Method for an improved traffic shaping and/or management of ip traffic in a packet processing system, telecommunications network, network node or network element, program and computer program product. European patent office application number: 20157452.2-1215. Deutsche Telekom AG. Feb 2020

18. Kundel R, Nobach L, Blendin J, Kolbe H-J, Schyguda G, Gurevich V, Koldehofe B, Steinmetz R (2019) P4-BNG: central office network functions on programmable packet pipelines. In: Proceedings of the 15th international conference on network and service management (CNSM). IEEE, Oct. 2019, pp 1–9
19. Kundel R, Nobach L, Blendin J, Maas W, Zimber A, Hans-Kolbe J, Schyguda G, Gurevich V, Hark R, Koldehofe B, Steinmetz R (2021) OpenBNG: Central office network functions on programmable data plane hardware. Int J Netw Manag 31(1):1–25
20. Kundel R, Nobach L, Kolbe H-J, Meuser T, Steinmetz R (2022) FPGA-assisted massive packet queueing and traffic shaping at the network edge. In: Proceedings of the 30th annual international symposium on field-programmable custom computing machines (FCCM). IEEE, p 1
21. Kundel R, Nobach L, Schneider F (2019) Method for an improved traffic shaping and/or management of ip traffic in a packet processing system, telecommunications network, system, program and computer program product. European patent office application number: 19215704.8-1215. Deutsche Telekom AG. Dec 2019
22. Kundel R, Rizk A, Blendin J, Koldehofe B, Hark R, Steinmetz R (2021) P4-CoDel: experiences on programmable data plane hardware. In: Proceedings of the IEEE international conference on communications (ICC). IEEE, pp 1–6
23. Kundel R, Rizk A, Koldehofe B (2020) Microbursts in software and hardware-based traffic load generation. In: Proceedings of the IEEE/IFIP network operations and management symposium (NOMS). IEEE, pp 1–2
24. Kundel R, Siegmund F, Blendin J, Rizk A, Koldehofe B (2020) P4STA: high performance packet timestamping with programmable packet processors. In: Proceedings of the IEEE/IFIP network operations and management symposium (NOMS). IEEE, pp 1–9
25. Kundel R, Siegmund F, Hark R, Rizk A, Koldehofe B (2022) Network testing utilizing programmable network hardware. In: IEEE communications magazine (Feb 2022), pp 12–17
26. Kundel R, Siegmund F, Koldehofe B (2019) How to measure the speed of light with programmable data plane hardware? In: Proceedings of the ACM/IEEE symposium on architectures for networking and communications systems (ANCS). IEEE, pp 1–2
27. Kundel R, Wallerich J, Maas W, Nobach L, Koldehofe B, Steinmetz R (2019) Queueing at the telco service edge: requirements, challenges and opportunities. In: Proceedings of the 1st workshop on buffer sizing. Stanford University, pp 1–6
28. Kundel R, Zimber A (2020) Verfahren zum Erfassen und Verarbeiten eines Zählsignals. European Patent Office Application Number: 20213595.0 - 1215. Deutsche Telekom AG
29. McKeown N, Anderson T, Balakrishnan H, Parulkar G, Peterson L, Rexford J, Shenker S, Turner J (2008) OpenFlow: enabling innovation in campus networks. ACM SIGCOMM Comput Commun Rev 38(2):69–74
30. Nielsen J (2019) Nielsen's law of internet bandwidth. https://www.nngroup.com/articles/law-of-bandwidth/. [Online; Accessed 26-January-2022]
31. Schmoll R-S, Pandi S, Braun PJ, Fitzek FH (2018) Demonstration of VR, AR offloading to mobile edge cloud for low latency 5G gaming application. In: 2018 15th IEEE annual consumer communications and networking conference (CCNC). IEEE, pp 1–3
32. Singh SK, Rothenberg CE, Patra G, Pongracz G (2019) Offloading virtual evolved packet gateway user plane functions to a programmable ASIC. In: Proceedings of the 1st ACM CoNEXT workshop on emerging in-network computing paradigms. ENCP '19. ACM, pp 9–14
33. Wang S, Wan J, Li D, Zhang C (2016) Implementing Smart factory of industrie 4.0: an outlook. Int. J. Distrib. Sens. Netw. 12(1)
34. Woesner H, Lange C, Schlenk R, Schlosser M, Kosiankowski D (2016) Virtualization in high-throughput network elements and its impact on energy consumption. In: Photonic networks; 17. ITG-symposium; Proceedings of 2016 May 12. VDE, pp 1–6
35. Zhang X, Chen H, Zhao Y, Ma Z, Xu Y, Huang H, Yin H, Wu DO (2019) Improving cloud gaming experience through mobile edge computing. IEEE Wirel Commun 26(4):178–183

Chapter 2
Background and State-of-the-Art

In this chapter, we provide background information as well as prior work in the field of this thesis. First, in Sect. 2.1, we introduce the general concepts of network softwarization. Section 2.2 gives an overview of the access technologies in residential and mobile Internet service creation. Following this, we summarize existing hardware acceleration technologies for computer networks in Sect. 2.3, which can be applied in Internet access networks. One focus of this work is on algorithms and mechanisms for Queueing and Scheduling at the Internet access edge. The belonging background and related work are presented in Sect. 2.4. Last, in Sect. 2.5, state-of-the-art approaches for network function benchmarking are presented.

2.1 Network Softwarization

Network softwarization is the primary enabler for rapid innovation in computer networks and includes the approaches Software Defined Networking (SDN) and Network Functions Virtualization (NFV).

Scott Shenker, one of the network softwarization pioneers, motivated SDN as follows: *"Think of it as a general language or an instruction set that lets me write a control program for the network rather than having to rewrite all of code on each individual router"* [82].

The Open Networking Foundation, a non-profit organization that standardizes multiple open network APIs, defines SDN as *"the physical separation of the network control plane from the forwarding plane, [...] where a control plane controls several devices"* [40]. Specifically, SDN approaches provide an open Application Programming Interface (API) to configure capable network devices by a logically separated controller [38, 50, 70]. The *data plane* is defined as the set of devices responsible for forwarding network packets, such as all network switches. All functionality not responsible for packet forwarding, e.g., routing engines, belongs to the *control plane*.

R. Kundel, *Accelerating Network Functions Using Reconfigurable Hardware*, Springer Theses, https://doi.org/10.1007/978-3-031-52872-9_2

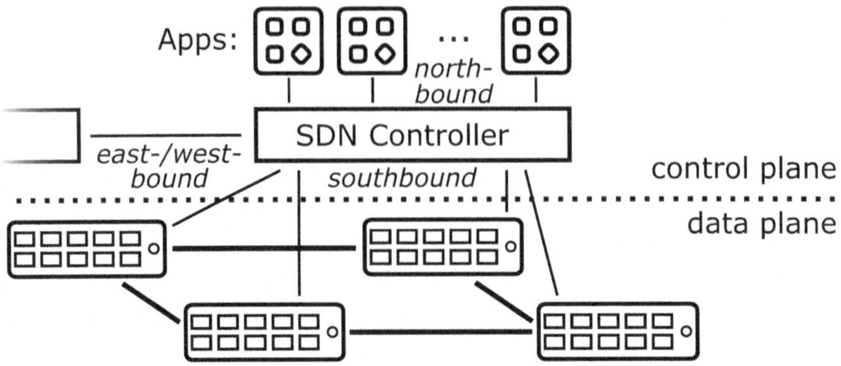

Fig. 2.1 Schematic structure of SDN-based networks

The most prominent implementation of such an API between data and control plane is *OpenFlow*, which was introduced in 2008 and supported by many commercial network switches [81]. An SDN controller holds a centralized global view of the network as a basis for its configuration decisions. Compared to a local and incomplete network view within traditional switches, better decisions can be performed, e.g., when routing of network flows. Further, the SDN controller can provide a second API to SDN Apps [100, 126]. This API allows disaggregating control functionality, whereas the controller offers only basic functionalities for event handling, monitoring, and flow rule installation. All further functionality, e.g., routing, DDoS detection, or network analytics, are executed as Apps. Consequently, the controller and all network Apps build the control plane. The *Open Network Operating System*, maintained by the Open Networking Foundation, is an open-source-available controller providing the introduced interfaces to network switches and SDN Apps [15].

As shown in Fig. 2.1, the interface between network switches and the controller is named *southbound*; the *northbound* interface describes the connection to the SDN Apps [8]. The east-/west-bound API allows interconnecting multiple SDN controllers for load balancing and reliability reasons. Note that multiple physical controller instances can form a single logical SDN controller with the complete network view.

However, the data plane functionality of the network switches remains unchanged. The internal behavior of SDN-capable data plane devices is typically realized by a fixed-function Application Specific Integrated Circuit (ASIC). To summarize, the flexibility in SDN is increased by opening up the configuration interface and decoupling and centralizing the network's control logic.

To overcome the limitations of fixed data planes, NFV was introduced in 2012 [25]. NFV is a concept for realizing data plane functionality as a software component running in commodity servers. By utilizing a server with a powerful Central Processing Unit (CPU) and a high-performance Network Interface Card (NIC), almost any network function can be realized flexibly, e.g., for IoT applications or within mobile access networks [5, 53].

The *Open vSwitch* is the most prominent implementation of an SDN-capable switch realized on commodity servers by utilizing NFV concepts [98]. This approach is already highly optimized and can achieve much higher data and packet rates than conventional Linux-based systems. The achievable throughput is mainly limited by the Dynamic Random Access Memory (DRAM) bandwidth of the system, i.e., DDR3 or DDR4 memory, to a few hundreds of Gbit/s [37]. For small average packet sizes, the packets rate denoted typically in *packets per second* (*pps*), is the limiting factor as for each packet a forwarding decision must be performed. However, the performance is at least a magnitude lower compared to ASIC-based networking hardware providing constant throughput even at small packet sizes [112].

To summarize, SDN and NFV provide concepts for describing networking functionality in a unified way. However, conventional SDN approaches still suffer under limited data plane functionality, and NFV does not provide similar performance.

2.2 Internet Service Creation

The main task of an ISP is the establishment and maintenance of connectivity for its subscribers. The network infrastructure of an ISP can be divided into multiple parts, as shown in Fig. 2.2.

The backbone network spans over a large geographical area, i.e., a nationwide network. Note that this network is sometimes also referred to as a *core network*; however, due to naming conflicts with the *5G core* we refer to it as the *backbone* network only. Many local access networks provide connectivity to this backbone via wired residential and wireless mobile access technologies. Betker et al. presented in 2014 a model of the network topology of a large ISP in Germany [16]. According to them, the core network consists of 9 core locations, each having two redundant core routers. Further, 900 residential access network nodes are connected to these

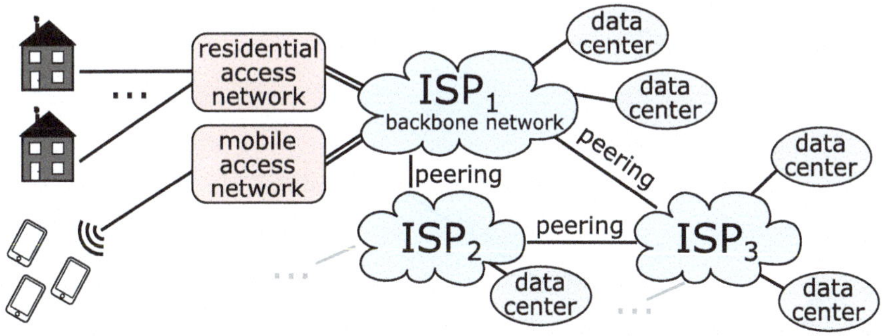

Fig. 2.2 Overall Internet architecture consisting of multiple ISP including mobile and residential access. Figure derived from: [73]

9 locations in redundant rings. Similarly, in the case of mobile access, an access network provides connectivity to the backbone network. The access networks of mobile and residential access can currently be considered to be independent but might converge in the future [125].

Besides subscribers, content providers, i.e., data centers, are connected to the network of an ISP. Large content providers typically have a worldwide content provisioning infrastructure being present in most operator networks [19]. However, some content, e.g., the webserver of a local sports club, might be present only in a single ISP network. For that, ISP networks can be interconnected with each other at *peering* points to make the resources in the counter-side network available.

Definition: The **Access Edge** describes all components and functionality required to provide Internet access for subscribers, excluding the subscriber equipment.

In the following subsections, we will describe in detail the structure of residential and mobile access networks, the *access edge* of the Internet.

2.2.1 Residential Access Networks

Residential access networks connect houses, apartments, and companies to the Internet or a private network. Due to the evolution of access networks over many years, multiple access technologies exist. Figure 2.3 presents common access technologies deployed in nowadays networks.

The Broadband Network Gateway (BNG) is responsible for terminating the subscriber lines of all attached Residential Gateways (RGs) and routing their traffic accordingly [54]. In older literature and specifications, the device realizing BNG functionality is named Broadband Remote Access Server (BRAS) [111]. In the following of this work, we use only the term BNG.

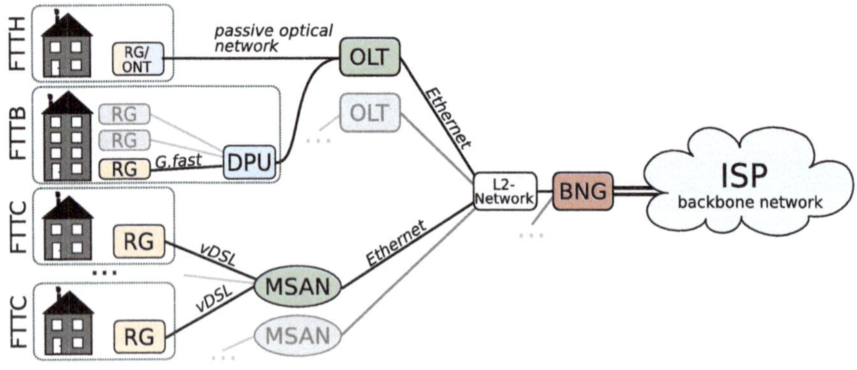

Fig. 2.3 Exemplary residential Internet access network topology. Figure derived from: [71]

The remaining access network does not perform any function on the network layer; it is only responsible for delivering data packets from the subscribers to the BNG and vice versa. The currently most common deployed technology is Very High-Speed Digital Subscriber Line (vDSL), connecting a subscriber via a copper wire to a Multi-Service Access Node (MSAN) [28, 32]. Starting from the MSAN, all subscriber data are sent in aggregated fiber-optical links to the BNG. This access technology is called Fiber to the Curb (FTTC).

While FTTC does not provide end-to-end connectivity over fiber-optical links, the more recent technology Fiber to the Home (FTTH) technology offers a high-speed fiber-optical connection between RG and BNG. Besides point-to-point connections, a passive optical network can connect n RGs to a single Optical Line Terminal (OLT) by optically splitting up the fiber as a shared, time-multiplexed medium [74]. This allows cost improvements as the OLT must provide fewer ports, and the available physical bandwidth of a fiber-optical cable seems to be no limitation at the moment and in the near future.

As the deployment of FTTH implies building activities in every building and apartment, it is not easy to upgrade existing houses. Therefore, Fiber to the Building (FTTB) was introduced as a technological compromise. A fiber-optical cable is installed into the basement of a building, and only the last meters to the RG are realized with conventional copper-based technology. Due to the shorter cable length, the signal-to-noise ratio is better, and subscribers can achieve higher data rates than with vDSL. This technology is named *G.fast* [122].

Besides these three technologies, other access methods exist and vary between countries. To summarize, a BNG must serve subscribers being connected by multiple access network technologies concurrently.

2.2.2 Functional Requirements of BNGs

In the following, we describe the functional data plane requirements of BNG systems. Note that these requirements vary from country to country and from operator to operator. Further, they provide room for interpretation, wherefore, even two operators in a country can have a different configuration of their network [119]. Based on existing literature and our previous work, we introduce the essential functionality of a BNG [4, 28, 71]:

Tunneling: For each subscriber, a tunnel between RG and BNG is established, e.g., by the use of the L2-Tunneling Protocol (L2TP), Point-to-Point Protocol over Ethernet (PPPoE) [24] or any other appropriate protocol. By that, subscribers are isolated from each other, and per-subscriber accounting can be performed. In the following of this work, we focus on PPPoE only, but the presented concepts can be generalized. The MSAN or OLT in Fig. 2.3 does not influence this tunnel as they only pass through all packets while converting from one physical medium to another. In total, for a single BNG 5,000–35,000 subscribers can be expected [105]. The maximum bandwidth of a BNG depends on its deployment scenario; however it can be assumed to be in the range of hundreds of Gbit/s. Last, an IPv4 (and an IPv6) address must be assigned to the RG of each subscriber.

Routing: The BNG must forward all packets to and from the subscriber. For that, a certain set of routes must be known and continuously synchronized with the backbone network. Further, the packet size may exceed the Maximum Transmission Unit (MTU) of the access network due to the additional PPPoE header, and either fragmentation or signaling is required [35]. In addition, the BNG decrements the *time to live* field of forwarded packets. If this field reaches zero, the sender must be notified. Last, for IPTV multicast services, the BNG must manage the multicast groups and duplicate all packets accordingly.

QoS and traffic shaping: A subscriber is allowed to utilize the access network up to a specific bandwidth, typically named in his Internet contract. To ensure this bandwidth is not exceeded but to allow the subscriber to utilize its bandwidth fully, queueing must be performed in the BNG for traffic towards the subscriber (downstream). For upstream traffic, i.e., packets sent from the subscriber to the Internet, a policing on the maximum rate is sufficient as the RG performs the queueing. The details on needs, queue dimensioning, and algorithms are presented later in Sect. 2.4.

Each subscriber can have different services belonging to numerous Quality of Service (QoS) classes, e.g., network management, Voice over IP (VoIP), multicast IPTV, and best-effort traffic. Packets belonging to a higher priority class must be able to overtake already enqueued packets with a lower priority. Consequently, a dedicated queue is required for all QoS classes of each subscriber. Assuming 35,000 subscribers and 4 QoS classes, 140,000 queues are needed. The configuration of the queues and schedulers offers a lot of freedom, leading to a different setup for each ISP, all having their pros and cons [119].

In order to achieve the often requested accounting accuracy, there must be zero packet loss between BNG and RG. Due to over-subscription, packet loss can occur within the access network. For example, an MSAN provides Internet connectivity to 400 subscribers, each having a 100 Mbit/s access line connected to the BNG only by a single 10 Gbit/s cable. However, it is very unlikely that all subscribers use their access line thoroughly at the same time. Nevertheless, this network node is over-subscribed by a factor of 4, and therefore packet loss can occur. To avoid this, in addition to normal QoS, the BNG must realize a Hierarchical Quality of Service (HQoS), which is aware of the hierarchical access network topology.

2.2.3 Softwarization of BNG Functionality

Realizing BNG functionality with SDN and NFV concepts has been discussed in the literature before.

Bifulco et al. presented a software-based architecture for BNG functionality following the NFV concepts in 2013 [17]. In contrast to conventional BNG systems, software realizations provide much higher flexibility as new instances can be simply started and stopped to serve the current bandwidth needs [61]. Further, subscriber sessions can be migrated at runtime from one instance to another to allow highly flexible scaling but also to handle a failover in case of a crashing software instance [34]. Subsequent works built upon this and presented the idea of considering a BNG as a usual data center [97]. However, these software-based approaches strongly suffer under limited bandwidth compared to ASIC-based solutions; e.g., Bifulco et al. named "up to 10 Gbit/s."

Until now, we considered the BNG as a single system. However, it can be disaggregated into a control plane and data plane, separated by a well-defined API, following the key idea of SDN [54]. By that, the complexity is decreased, and innovation within one of these components is simplified. Utilizing SDN protocols for managing access network nodes, e.g., the MSAN, has been investigated in literature as well [65].

Nobach et al. investigated the realization of BNG-functionality with OpenFlow-capable SDN switches. However, they figured out that the OpenFlow protocol does not support all required functionality, mainly the processing of PPPoE packet headers. Further, even if an extension of the protocol would allow the description of the needed functionality, the networking hardware still does not implement these features, a general limitation of pure SDN approaches.

2.2.4 5G Mobile Access Networks

Access networks for mobile Internet service creation are specified by the 3rd Generation Partnership Project (3GPP). In this work, we focus on the 5G standalone standard only, specified in *Release 15* of the 3GPP [2]. Previous 5G standards of the 3GPP specify "5G non-standalone" only. The non-standalone standard builds upon a 4G-Core and two radio cells for connecting the User Equipment (UE) with the network. Building upon 4G technology only, the first cell is used for all control functionality, including dial-in and data transmission. This cell is called "anchor cell." The signaling methods are very similar to conventional 4G networks. The second cell, building upon 5G technology, is used as a second data channel to the UE in order to increase the total bandwidth. This 5G radio channel is activated only if needed to reduce the UE energy consumption. As this works focus on data plane functionality within the core network, we only focus on the standalone architecture.

Figure 2.4 depicts the end-to-end connectivity of UEs in 5G standalone access networks. Generally, the access network can be divided into two functional parts: the Radio Access Network (RAN) and the 5G core. Further, ISPs deploy a Network Address Translation (NAT) between the mobile access network and their backbone network to hide the public IP addresses of their subscribers for security reasons but also to save scarce public IPv4 addresses [79]. Following the SDN terminology, the architecture can be divided into a data plane (indicated by thick lines) and a control plane (indicated by thin lines).

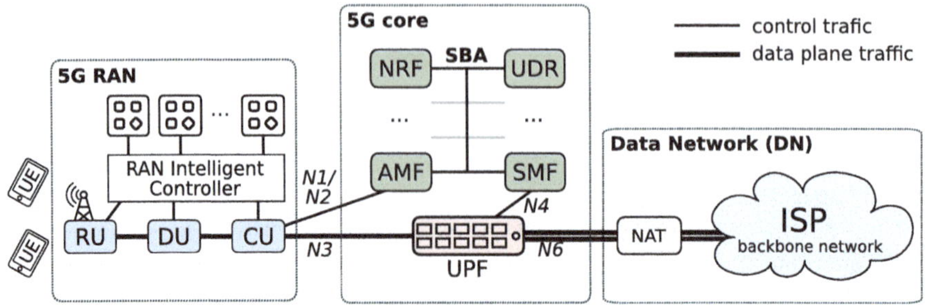

Fig. 2.4 Disaggregated 5G standalone architecture according to [3]

2.2.4.1 5G Radio Access Network

The RAN functionality can be further disaggregated into multiple network functions [78]. Following the *O-RAN* terminology, we present the RAN as a composition of the Radio Unit (RU), Distributed Unit (DU), and Centralized Unit (CU) disaggregated by a "7.2-split" [7, 43]. Further, an SDN controller, called *RAN Intelligent Controller (RIC)*, and SDN Apps can (but most not be) deployed. This work focuses on the data plane acceleration, wherefore we describe these three network functions in more detail. Note that the data plane is often named *user plane* in the context of mobile networks; however, we use only the term *data plane* in the following for reasons of uniformity.

The radio unit contains mainly high-frequency functionality, i.e., sending and receiving radio signals. The main functionality, e.g., scheduling, pacing, baseband modulation, encryption, is performed within the distributed and centralized unit. Due to this functionality, the functionality in the data plane of the RAN is very compute-intensive and can enormously benefit from hardware acceleration [18, 20]. Further, the centralized unit contains basic control plane functionality for session management, which is less performance-demanding than the data plane.

2.2.4.2 5G Core

The 5G core is a compound of multiple network functions in a Service-Based Architecture (SBA). Besides the User Plane Function (UPF), all functions belong to the control plane, e.g., the NRF, UDR, AMF, and SMF, and are not the focus of this work. However, we described the setup used in this work in Appendix A.2.

The UPF is responsible for the termination of all subscriber sessions, analogous to the BNG in residential access networks [3]. In contrast to the BNG, all control functionality is realized within the control plane functions of the 5G core, and the UPF can focus on subscriber traffic processing. Traffic is encapsulated between the RAN and the UPF with the GPRS Tunneling Protocol (GTP) protocol for the same reasons as in residential access networks [1]. Therefore the UPF must perform this encapsulation based on flow rules installed by the 5G core control plane. Further, the UPF must perform a QoS-aware packet queueing. Analogous to residential Internet access, several QoS classes for VoIP, best-effort traffic, and other services may exist.

2.3 Programmable Hardware for Computer Networks

The use of hardware accelerators, including programmable hardware, can enormously improve the overall system performance in computer networks and has been the subject of many previous research works and surveys [52, 66]. In this work, we use the following definition:

Definition: A programmable **Hardware Accelerator** executes a function more efficiently by utilizing specialized, reconfigurable hardware components, e.g., an Field Programmable Gate Array (FPGA), a Graphics Processing Unit (GPU), a Network Processing Unit (NPU), or a programmable switching ASIC.

This means that in contrast to the original NFV approach, introduced in Sect. 2.1, commodity CPUs are partially or fully replaced by hardware accelerators. In the following subsections, we introduce multiple hardware accelerator technologies being suitable for improving the performance of network functions.

2.3.1 Programmable Network Switches

Fixed-function ASICs in networking hardware do not support any reconfiguration exceeding the chip functionality at runtime. Besides controlling table entries, only interface parameters, e.g., link speed and auto-negotiation, and basic traffic shaping policies, can be set. Special network protocols and processing sequences are either supported or not, e.g., the PPPoE protocol in residential access networks. This limitation also exists in conventional SDN hardware, i.e., OpenFlow-capable switches.

To overcome this, Bosshart et al. presented in 2013 a general and programmable switch pipeline architecture [22]. Their approach, presented in Fig. 2.5, consists of three main components: (1) After packets ingress the switch on one of the n ingress ports, the ingress pipeline processes the packet first. (2) Next, the packets are stored in configurable output queues and scheduled according to the egress port's link speed and configured rate. (3) Last, after being scheduled, the packets traverse the egress pipeline.

The internal design of the ingress and egress pipeline is the same and consists of a fixed number of programmable pipeline stages. Each pipeline stage consists of a generic set of resources, which can be configured and interconnected at runtime. The parsed packet header of a packet and its meta-data traverse the pipeline from the left to the right and the pipeline cannot be stalled. Therefore, each pipeline stage must

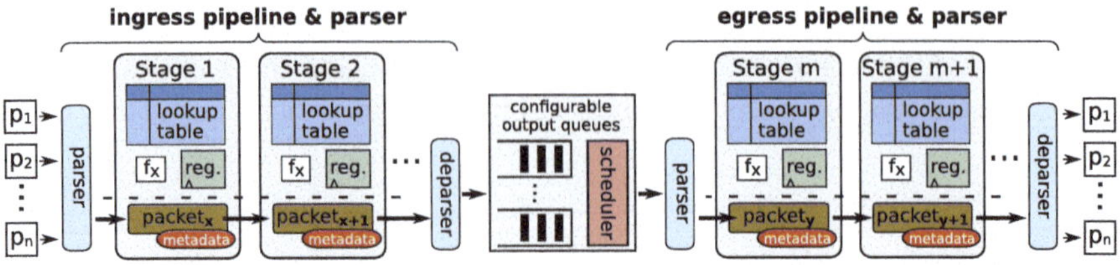

Fig. 2.5 Generic programmable packet processing pipeline following the P4 programming paradigm. Figure derived from: [22, 72]. © 2021 IEEE. Reprinted, with permission, from "P4-CoDel: Experiences on Programmable Data Plane Hardware" [72]

have a guaranteed and bounded packet processing time. Meta-data are information belonging to the packet but not being part of the packet, e.g., the ingress port, ingress time, queueing delay, or computed values.

The resources in a pipeline stage include memory blocks that can be utilized for *lookup tables* to perform a lookup on one or multiple input values, e.g., a packet header field or a computed value. Further, stateful operations can be performed in *registers* to maintain a state between the processing of two packets. Last, arithmetic operations, denoted as f_x, can be performed on any packet header or meta-data field.

Before a packet enters the pipeline, it will be parsed by the *parser*. The developer can define a parser state machine to extract any packet header format. Next, the parser provides a vector of parsed header fields to the pipeline. All unparsed bytes of a packet are defined as *payload* and bypass the pipeline. Similarly, at the end of the pipeline, the *deparser* serializes the header vector, possibly modified within the pipeline, and attaches the *payload* to it.

In order to program this generic structure, the Programming Protocol-Independent Packet Processors (P4) language was introduced in 2014 [21]. P4 is a C-like language optimized for describing the behavior of data planes and can be compiled for generic hardware architectures. An example code for a simple Ethernet switch is presented in Sect. A.1. The P4 language released in 2014 is restricted on this pipeline architecture consisting of an ingress and egress pipeline, wherefore other switches and SmartNICs could be only programmed with strong limitations. Consequently, $P4_{16}$ was presented in 2016 to overcome these limitations and enable further extensibility [23].

To the best of our knowledge, currently, two ASICs exist following the concept of general and configurable hardware: The Intel Tofino and the Pensando DSC [58, 96]. However, further P4-programmable platforms, not following such an ASIC architecture, exist, i.e., FPGAs and NPUs, which will be presented in the following subsections.

P4-programmable ASICs provide an alternative to CPUs in NFV, e.g., for offloading mobile access network functionality [112]. Besides programmable packet processing, programmable hardware for optical circuit switching exists, which can be used to improve the flexibility in softwarized networks but is not the focus of this work [131].

2.3.2 Field Programmable Gate Arrays (FPGAs)

Field Programmable Gate Arrays (FPGAs) are programmable chips allowing to realize any boolean logic by reconfiguration only. They can be programmed by hardware description languages, e.g., Verilog and VHDL [56, 57]. The functionality of any ASIC can be realized with FPGAs, assuming sufficient resources are available. By using FPGAs, the development costs can be strongly reduced, and the innovation cycle time is lower. Further, the configuration of an FPGA can be easily changed after manufacturing, allowing new product features and bug fixes while being in operation [123]. Compared to P4-programmable ASICs, introduced in the previous subsection, the expressiveness on FPGAs is much higher. However, FPGAs suffer under a lower performance due to their clock frequency, approximately a quarter of ASICs. Consequently, the maximum achievable throughput is lower, and FPGAs should only be used when programmable ASICs do not provide the required functionality. This guidance is strengthened by the fact that developing algorithms on FPGAs at bit-level is much more time-intensive than describing data plane behavior in P4.

Utilizing FPGAs for networking functionality has been discussed in related work several times. The *NetFPGA* platform, introduced in 2006, provides a common base for developing network functions in a framework tailored for network applications with less effort [129]. In the meantime, further commercial and non-commercial frameworks exist, and many networking applications have been presented, e.g.: Nagy et al. proposed an FPGA accelerated DDoS attack detection system which can make a decision within milliseconds on the FPGA [85]. Further, the in-network processing of video streams in QoS-aware 5G networks has been demonstrated by Ricart et al. [104].

Besides networking, FPGAs are used in many other domains: They can be used to strongly accelerate encryption methods, for example *OpenSSL* [67]. They are also very prominent in big data evaluations and many other fields, which require high performance, where building ASICs is not reasonable [86].

To overcome the hurdle of high development effort, multiple projects aimed at a P4 to FPGA compiler [14, 127]. Despite mapping the P4 language constructs directly onto the available hardware resources, all known approaches build upon an intermediate representation in a low-level FPGA programming language, e.g., Verilog, VHDL, or Bluespec. Currently, two commercial compilers from Xilinx and Intel, the world's largest FPGA vendors, are available and can be used to describe a design being composed of a P4 pipeline and low-level modules for specialized functionality [14, 90]. On top of such a compiler, the "*P4 → NetFPGA*" project was created, providing the complete workflow to compile a P4 program on FPGAs without any low-level FPGA programming [55]. Consequently, P4 compilers provide a tool to establish FPGAs without requiring FPGA developers.

2.3.3 Graphics Processing Units (GPUs)

Graphics Processing Units (GPUs) are well known for personal computers, where they are mainly used to render images, which are displayed on the screen. Internally, they consist of many processing cores optimized for vectorized arithmetic. Although these linear algebra processors were only intended for graphics computation, they can be used in many other application scenarios, e.g., machine learning, scientific computing, and networking function acceleration. Network Functions can typically process many packets in parallel and, therefore, are very well suited for an acceleration in GPUs [63, 118, 124]. In contrast to FPGAs, they are more commonly used, and the price per device is lower. However, due to the nature of GPUs, they do not provide network interfaces, and all data is copied into and from the GPU by a helper process running on the CPU of the host system. These copy processes cause a significant CPU and data bus utilization.

Internally, GPUs consist of many cores, typically more than 1,000. Only programs that can be executed parallelly benefit from this architecture, as a single GPU-core has less performance than a server or desktop computer CPU. In contrast to conventional multi-core CPUs, the GPU architecture is optimized for executing the same program code on many cores in parallel, known as Single Instruction Multiple Data (SIMD) parallelism. However, modern GPUs can run numerous *streams* in parallel, each consisting of a SIMD-program, providing additional flexibility [115]. The expressiveness and flexibility of programs written for GPUs are similar to conventional software programs running on CPUs and, therefore, much higher than on P4-programmable switches. GPUs can be programmed by domain-specific languages, e.g., OpenCL or CUDA [47, 91].

2.3.4 Smart Network Interface Cards (SmartNICs)

Smart Network Interface Cards (SmartNICs) is an extensively used term and includes various technologies. The common feature of all SmartNICs is that they have the same form factor as an ordinary Network Interface Card (NIC) but additionally offer special functions. This function set can include advanced load balancing, security features, remote direct memory access [49], or programmability of the data plane.

Various chips can be manufactured on Smart Network Interface Cards (SmartNICs), including Field Programmable Gate Arrays (FPGAs), Network Processing Units (NPUs), and purpose build ASICs. Harkous et al. presented a comparison on different P4-programmable hardware, including two SmartNICs and one programmable switch [51]. While the investigated chipset of the P4-programmable switch showed very constant performance characteristics, the NPU- and FPGA-based SmartNIC implementations pose an increased latency for more complex data plane programs. The internal architecture of these chipsets causes this, being less performance-optimized than switching ASICs with many ports and $\geq$10 Tbit/s total bandwidth capacity. As the required amount of SmartNICs is high, the chipset price must be as low as possible and the ASICs from programmable switches, providing an over-dimensioned performance, is uneconomic. An NPU is a processor specially built for performing network packet processing, e.g., the *Netronome NFP* SmartNICs build upon an NPU [88]. However, a new chipset was announced by *Pensando, Inc.*, claiming to have a switch-like internal pipeline providing constant and deterministic behavior within the SmartNIC [42]. As of today, no evaluation results exist for this architecture, either fortifying or falsifying this statement. In addition, the GPU vendor *NVIDIA* announced SmartNICs, called *Bluefield*, utilizing a hardware architecture similar to GPUs [30].

In the previous Sect. 2.1, we introduced the Open vSwitch architecture for pure software-based packet switching. By utilizing SmartNICs with specialized hardware, i.e., an FPGA or an NPU, the performance of this software switch can be increased by a factor of two to four [87, 130].

2.3.5 Control Plane Interfaces

The control plane interface of networking hardware must provide the required functionality to configure the underlying hardware. As OpenFlow switches provide a fixed and specified function set, the control plane API can be static [81].

However, programmable data planes allow allow the description of any control flow and table format, wherefore, more generalized APIs are required. The P4 programming language can be used to describe the behavior of the data plane, and a compiler generates a hardware configuration file out of it. Besides this, the compiler can generate a custom API tailored solely for this P4 program, called *P4Runtime* [92]. This approach can even be used to describe the behavior of fixed-function ASICs as a P4-model in order to auto-generate an API for this architecture. As one of the goals of the P4 programming language is hardware independence, a single P4 program can be compiled to multiple hardware targets, all providing the same P4Runtime API. The P4Runtime API builds upon the google Remote Procedure Call (gRPC) protocol [48]. In case the behavior of the data plane exceeds the expressiveness of P4, a custom API must be defined, e.g., utilizing gRPC, REST, or any other suitable protocol technology.

2.4 Scheduling and Active Queue Management

Human users of the Internet expect a good subjective performance of the network, known as Quality of Experience (QoE) [59]. However, this metric depends strongly on the perception of test persons, and it takes high effort to quantify and measure it accurately, e.g., with user studies.

Therefore, Quality of Service (QoS) metrics were introduced to assess systems without human test subjects [80, 107, 117]. Further, machinery systems with an underlying network, e.g., within smart factories, require a certain QoS level [44, 45]. In this work, we use the following QoS definition:

Definition: "Quality of Service (QoS) is the well-defined and controllable behavior of a system with respect to quantitative parameters" [107].

The important characteristic of QoS metrics is the measurability of quantitative parameters. In the context of computer networks, mainly the following metrics are considered: latency, throughput, latency jitter, rate jitter, prioritization, packet loss, availability, and out-of-order delivery of packets [80].

Latency describes the duration a network packet traverses the network from the sender to the receiver. The maximum *throughput* of this connection is typically limited by the bottleneck link with the lowest available bandwidth on the path. *Jitter* can be subclassified in *latency jitter* and *rate jitter*, both being indicators for a non-deterministic network performance, e.g., caused by a single overloaded network function. A variance in the latency of multiple packets transmitted on the same route is named *latency jitter*. *Rate jitter* is characterized by a varying throughput over time. *Prioritization* describes if packets are forwarded according to their QoS class, i.e., high-priority packets can overtake packets with lower priority in case of congestion within a network switch. The metric *packet loss* describes the number of packets not delivered to the receiver, typically presented in relation to the total amount of sent packets. The relative downtime of a network is defined as *availability*. *Out-of-order* packet delivery can occur in networks if the latency jitter between two packets is higher than the inter-packet time. Note that out-of-order delivery between multiple traffic classes can occur due to prioritization and is not always negative. Also, packet reordering is not necessarily a problem within a traffic class but can often cause reduced end-to-end performance [128].

2.4.1 Queueing and Congestion Control in Access Networks

Queueing is required to achieve high throughput in nowadays computer networks, especially in network switches with a lower egress than ingress bandwidth.

Typically transport protocols in computer networks use a congestion control mechanism to avoid sending more data than the network can deliver to the receiver. Due to the dynamics of many flows in computer networks, this congestion control constantly updates its sending rate based on network feedback [95]. The most common feedback is packet loss, detected by the receiver and cascaded to the sender by not acknowledging the packet. Further, a congested network switch can label a packet with an "explicit congestion notification" marker to notify the receiver and, by that, the sender. Both mechanisms result in a reduced sending rate and consequently avoid congestion.

As the available bandwidth in the network can vary, the congestion control increases its rate until the first packet loss occurs. In case of packet loss, it will reduce its rate significantly and continues increasing the rate, also known as *congestion window*, which describes the maximum number of packets in transmission. By that, the sending rate is oscillating around the maximum achievable throughput. To utilize a bottleneck link fully by one or multiple congested controlled flows, a queue before this link is needed, shaping the packets on link rate. The optimal queue size depends on many factors, including the number of congestion-controlled flows, the link speed, the flow Round-Trip Time (RTT), and the congestion control mechanism. On the one hand, the link will not be fully utilized if the queue size is too small. On the other hand, if the queue size is too big, unnecessarily high latency will be caused by queueing the packets. According to Appenzeller et al., the optimal queue size for TCP flows is [10]:

$$B = \frac{\overline{RTT} \cdot C}{\sqrt{n}}$$

Where $\overline{RTT}$ denotes the average RTT over all flows, C the bottleneck link speed, and n the number of parallel congestion-controlled flows. This formula was derived for Transmission Control Protocol (TCP), the most used transport protocol in the Internet, but other protocols utilize similar congestion control mechanisms, i.e., the QUIC transport protocol [73, 75]. Nowadays, the access network is typically the bottleneck link, and ~93% of the residential Internet traffic is congestion-controlled. Therefore, queueing at the access edge is of tremendous importance to ensure a good link utilization and will be discussed in this work [73].

2.4.2 Queueing Theory

This section introduces the terminology of queueing systems we will use in this work. Figure 2.6 depicts the general setup of a queueing system. In the first step, packets ingress on the left side into the system, and the according queue will be determined by a *classifier*. If the queue reaches its maximum capacity, packets are dropped. This procedure is called "taildrop." The system can consist of one or multiple First In First Out (FIFO) queues. On the right side, a *scheduler* determines which packet should be sent next. This scheduler is aware of rate limits of the individual queues, of different QoS classes, and all other information to be considered, e.g., hierarchical rate limits in residential Internet access networks. Suppose different QoS classes are assigned to the queues. In that case, several mechanisms exist to prioritize the scheduling process, e.g., Round Robin (RR), Weighted Round Robin (WRR), Weighted Fair Queueing (WFQ), Strict Priority (SP), and many others [121, Sect. 5.4.3]. Further, the scheduler can contain an Active Queue Management (AQM) algorithm, which decides to drop a packet instead of sending it to notify the congestion control. These

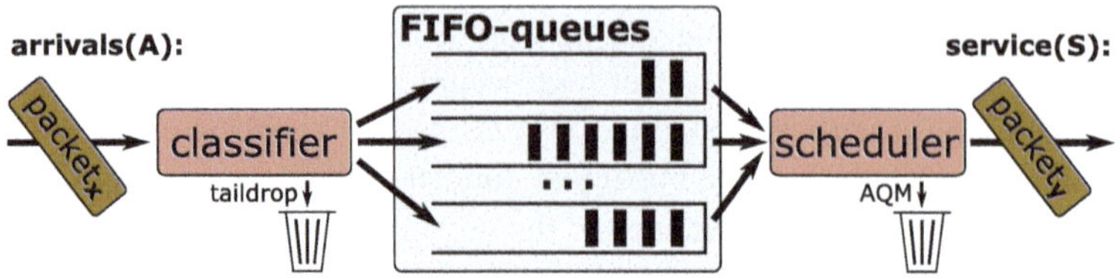

Fig. 2.6 Generic Queueing System Model

AQM algorithms are described in detail in the following two subsections of this work. Last, the packet will be sent by the egress port.

Queueing theory has a very long history, e.g., considering a system of humans waiting in a line to be served. We can also apply these theories to formalize the previously introduced packet queueing system. In the following, we build upon the model described by *David George Kendal* in the 1950s [64]. In general, the *Kendall notation* is denoted as follows [29, 69]:

$$A/S/c/B/N/SD$$

A describes the distribution of the arriving packets and S the distribution on the service side. In the case of networking hardware, we can assume the arrival process to be randomly distributed, indicated in the Kendall notation as $A = M(Markovian)$. If one egress port is sending packets from a single queue with a constant bit rate, e.g., 100 Mbit/s, the service is deterministic and therefore $S = D(Deterministic)$. However, in the case of multiple queues for one egress port, i.e., various QoS classes exist, we cannot assume the queue service to be deterministic and therefore $S = M(Markovian)$. Nevertheless, in practice, this behavior would be close to the deterministic queue as long as no congestion of the egress port occurs. The symbol c indicates how many service nodes are pulling packets from the queue, in the case of packet schedulers $c = 1$, as one egress queue is served by exactly one egress port. B describes the maximum number of entities in a queue. We determine this to be $B = B_{max}$, i.e., the queue capacity in bytes. The total amount of packets in the system is described by the symbol N: In the case of computer networks, we assume this to be $B = \infty$ as no upper bound can be given. Last, the service discipline SD describes in which order packets in the queue are processed. In this work, we focus on FIFO queues only; however, further disciplines exist. Therefore, we can describe a packet queueing system with a single queue per egress port as:

$$M/D/1/B_{max}/\infty/FIFO$$

and a system with multiple queues per egress port is described by:

$$M/M/1/B_{max}/\infty/FIFO$$

Upon these initial queueing models, the framework "Network Calculus" for describing and understanding the behavior of computer networks has been created [39, 76]. In the following of this work, especially the concepts presented in Chap. 4, we build upon these mathematical models.

2.4.3 Bufferbloat and Active Queue Management AQM

The phenomenon *bufferbloat* was first mentioned in 2011 and describes unnecessarily high latency in computer networks caused by generously dimensioned packet queues (buffers) in network switches. Bufferbloat can occur in many different kinds of networks, including residential and mobile Internet access networks in the downstream and upstream direction, as long as queues before bottleneck links exist [60]. As a consequence, a very high end-to-end latency can enormously decrease the QoS and by that the experienced service quality. As presented before, the optimal buffer size depends on multiple variables, and therefore addressing this issue is not trivial. Especially, the varying flow-RTTs in the Internet poses a hurdle.

Active Queue Management (AQM) algorithms can help to overcome issues regarding TCP flow synchronization and the bufferbloat phenomenon. In general, an AQM is "a method that allows network devices to control the queue length or the mean time that a packet spends in a queue" [13]. This control is performed by selective dropping or marking packets within the congested network device. In this work, we define them as follows:

Definition: Active Queue Management (AQM) algorithms manage the length of a packet queue by intended dropping or marking packets which is noticed by the transport-layer congestion control.

The first AQM algorithm, presented in 1998, is Random Early Detection (RED). This algorithm tackles the issue of synchronization effects when a taildrop queue is used, tending to overflow from one moment to the next, and massive packet loss occurs [27]. This massive packet loss affects all congested control flows simultaneously, wherefore they synchronously reduce their sending rate and become synchronized. In contrast to this, RED starts dropping packets with dynamically computed probability as soon as a threshold is reached, e.g., 60%, and this probability increases with the fill level of the queue. Consequently, only a few packets are dropped, congestion control mechanisms can react earlier, and no synchronization effects occur.

However, the RED algorithm does not solve the bufferbloat problem as it still causes a statically configured queue delay. To that end, more advanced and stateful algorithms were presented in the last decade:

The Controlled Delay (CoDel) and Proportional Integral controller Enhanced (PIE) algorithms avoid bufferbloat by allowing only a temporary high queue latency [89, 93]. Specifically, these algorithms maintain a state since when the queue latency exceeds a given threshold. After a waiting period in this exceeded state, they start dropping or marking packets. On the one hand, flows with a high RTT requiring large queueing delay can temporarily fill the queue to achieve high throughput. On the other hand, flows with a short RTT or many parallel flows cause a constant high latency, and the AQM would intervene. The justification of AQM algorithms for combining high throughput and low latency has been discussed multiple times in related work [46, 108]. Besides the presented algorithms, further variations and more specialized algorithms exist as well [33].

2.4.4 Hardware Realizations of Qeueueing and Scheduling

As network switches are typically based on ASICs, their behavior cannot be adopted easily and, as of today, mainly provide only basic taildrop and RED functionality. The previously presented approaches of programmable network switches, i.e., by the programming language P4, focus on header processing only, and the queueing functionality is still realized with a fixed behavior.

Sivaraman et al. presented the idea of utilizing FPGAs for scheduling, as they can be programmed very flexible, and the required algorithm varies from use case to use case [114]. In follow-up work, they introduced the concept of programmable schedulers by dynamically composing basic building blocks, i.e., Push In First Out (PIFO) queues, arbiters, and token bucket instances [113].

Although the PIFO architecture provides high flexibility, it has neither academic nor industrial success so far, possibly as a hardware realization seems to be challenging. *Shrivastav* proposed in a subsequent work the concepts of Push In Extract Out (PIEO) queues to improve flexibility and increase performance simultaneously [110]. Upon a similar approach, Zhang et al. investigated the capabilities of programmable queues in industrial networks with real-time guarantees [132].

Another research team proposed an approximation of PIFO queues by utilizing multiple FIFO queues [6]. Sharma et al. presented a concept for realizing calendar queues, a flexible data structure to realize improved schedulers, in P4-programmable ASICs [109]. However, this concept is not yet realized by an existing chip, and the initial realization causes the high costs and development times of ASICs. To summarize the related works, many queueing mechanisms for multiple scenarios exist; however, a realization with good performance in networking hardware is still an open challenge.

2.5 Benchmarking of Network Functions

An improved QoS in computer networks and improved architectures within servers for low-latency I/O operations can enable new application fields, e.g., virtualizing 5G radio functionality [43, 106]. However, not meeting the desired QoS level in computer networks leads to a lowered service quality and possible system failures; for example, in 5G fronthaul networks or production plants that rely on real-time communication [99, 103, 106]. In the context of 5G access networks, a guaranteed network QoS is often named "Ultra Reliable and Low Latency Communication (URLLC)", specifying a guaranteed latency up to <1 ms RTT and a reliability $>99.9999\%$ [68, 77]. For ultra-low latency applications, e.g., high-frequency stock trading, network switches exist which do not store a packet entirely before forwarding it and therefore have a port-to-port latency of fewer than 100 ns [11]. Therefore, it is important to ensure and validate the translation of the QoS metrics mentioned in the previous section.

As the testing methodology of network functions, black-box testing has been established as the most common way of testing [84]. This approach does not consider the internals of the tested network function, and only the interaction with the surrounding world is observed. In the case of a network function, this includes observing (1) the configuration of the data plane and (2) all ingressing and egressing packets. In general, the latency of modern network switches is very low, i.e., typically in the range of hundreds of nanoseconds, and the throughput is very high. Therefore, accurate measurement equipment, providing at least the same performance as the tested network function, is needed [101].

However, measuring such network functions is very challenging. Utilizing software-based tools for debugging or monitoring computer networks neither provides the required throughput performance nor the accuracy, so hardware assistance is mandatory [12]. Ideally, a test tool can generate small packet sizes up to the link speed, i.e., 100-byte packets at 100 Gbit/s, while providing a high latency and packet loss accuracy, i.e., nanosecond granularity.

2.5.1 Existing Approaches for Network Function Benchmarking

Multiple concepts for network function benchmarking, including commercial products, and academic projects, exist and will be presented in this section.

Cisco and Juniper, two large network suppliers, provide freely available software load generators, named *TRex* and *Warp17*, for load generation with high performance and traffic patterns up to the application layer [26, 62]. Both approaches utilize servers with state-of-the-art Network Interface Cards (NICs) and the Data Plane Development Kit (DPDK). DPDK allows direct control of the NIC without any operating system interference and, therefore, improves the performance compared to conventional software [41]. However, the accuracy of time measurements has only a granularity of around 100 ns and can be even worse under load [102]. To overcome this, the *Moongen* load generator, presented by Emmerich et al., utilizes the same NICs but with enabled hardware timestamping, initially intended for time synchronization [36]. By that, packets can be timestamped at ingressing or egressing the NIC with much higher accuracy, up to 10 ns [102]. However, this high accuracy can only be achieved when either the total load is low or a subset of packets is measured. Otherwise, small packet queues can be built up within the NIC. In all aforementioned software-based approaches, packet loss detection is difficult as no guarantees for a non-overloaded detection system can be given. This means detecting zero loss with one of these three generators is a correct result; however, the reason for packet loss cannot be determined for sure.

Solutions building upon programmable hardware, mainly FPGAs, exist to overcome these limitations. Micheel et al. presented the initial version of the *Data Acquisition and Generation* card in 2001 [83]. These cards are built upon an FPGA with a special-purpose design for load generation and packet capturing. They have the same form factor as commodity NICs. Current versions provide up to four 10 Gbit/s Ethernet ports, and large amounts of data can be stored directly in the hosting server's memory. The achievable accuracy and performance are slightly higher than with the aforementioned *Moongen* approach. However, they provide only limited flexibility, and due to the lower clock frequency of FPGAs, the maximum possible time accuracy is limited. Note that one clock cycle on an FPGA, running at 200 MHz, is 5 ns, limiting the measurement granularity. ASICs, running typically with at least 1 GHz, could provide a sub-nanosecond granularity but are far too expensive to develop for this niche market only. Besides this commercial product, further open-source projects for load generation within FPGAs exist [9, 31]. Other commercial test tools exist as well, i.e., from IXIA and Spirent, which suffer from very high prices and limited flexibility [36]. The internals of these solutions are not publicly available; however, we assume that they build upon FPGAs and could partially confirm this by datasheets [116].

Last, we would like to mention the In-band Network Telemetry (INT) approach presented in parallel to this work [120]. Instead of network function benchmarking, the purpose of INT is fine-grained monitoring in production networks. This approach leverages programmable network switches to monitor a single packet flow through the network. For this, at the ingress node, an INT header is added to the packet, and each switch on the path can add additional monitoring data to this header stack. This header stack is removed at the egress node, and the initial packet leaves the network. Further, the header stack of collected information can be exported at the egress node. The INT approach builds upon the same programmable switches we utilize in this work, and thus the theoretically achievable time accuracy can be assumed to be similar.

2.5.2 Measurement Data Acquisition

A considerable challenge in benchmarking network functions is the capturing of information. Let's assume a packet source of 10 Mpps, and each packet should be timestamped and classified for loss detection. If the investigated network function causes 1 ms latency, there are always 10,000*packets* in flight. For each of these packets, a state must be maintained in the test equipment, and deleted after a certain timeout in case of packet loss. In the case of previously discussed *Data Acquisition* *and Generation* cards, this is done within the internal memory of the card, which is realized with high-speed but rare SRAM memory, which can quickly become a limitation [94]. Further, a time series of all traversed packets must be stored persistently, which is impossible within the FPGA due to limited memory capacity. For that, the *Data Acquisition and Generation* cards can write the data directly via DMA into the host memory.

Another approach of measurement data acquisition is aggregating statistics directly in the data plane. Network switches can generate statistics such as transferred packets and bytes for each flow, and by periodic retrieving of counter registers, a time series can be generated [133]. For example, a sampling rate of 1 ms provides fine-grained results of throughput and packet loss; however, no information can be determined on a per-packet basis, e.g., latency distributions.

We summarize that multiple approaches for network function benchmarking exist. However, providing a flexible test solution while providing nanosecond accuracy at high data rates is challenging, and to the best of our knowledge, no existing approach can provide this concurrently.

References

1. 3GPP (2019) GPRS Tunnelling Protocol (GTP) across the Gn and Gp interface. Technical report (TR). Version 15.5.0
2. 3GPP (2019) Release 15 description; summary of Rel-15 work items. Technical report (TR). Version 15.0.0
3. 3GPP (2020) System architecture for the 5G System (5GS). Technical report (TR). Version 15.12.0
4. Agilent Technologies, Inc. (2006) Understanding DSLAM and BRAS access devices. White Paper
5. Alam I, Sharif K, Li F, Latif Z, Karim MdM, Biswas S, Nour B, Wang Y (2020) A survey of network virtualization techniques for Internet of Things using SDN and NFV. ACM Comput Surv (CSUR) 53(2):1–40
6. Alcoz AG, Dietmüller A, Vanbever L (2020) SP-PIFO: approximating push-in first-out behaviors using strict-priority queues. In: 17th USENIX symposium on networked systems design and implementation (NSDI 20). USENIX Association, Santa Clara, CA, pp 59–76
7. O-RAN Alliance (2021) O-RAN minimum viable plan and acceleration towards commercialization. White Paper
8. Amin R, Reisslein M, Shah N (2018) Hybrid SDN networks: a survey of existing approaches. IEEE Commun Surv Tutor 20(4):3259–3306
9. Antichi G, Shahbaz M, Geng Y, Zilberman N, Covington A, Bruyere M, McKeown N, Feamster N, Felderman B, Blott M, Moore AW, Owezarski P (2014) OSNT: open source network tester. IEEE Netw 28(5):6–12
10. Appenzeller G, Keslassy I, McKeown N (2004) Sizing router buffers. ACM SIGCOMM Comput Commun Rev 34(4):281–292
11. Arista Networks, Inc. (2021) Arista networks launches sub 100ns ultra-low latency switch for financial services. Press Release. https://www.arista.com/en/company/news/press-release/12604-pr-20210518
12. Arlos P, Fiedler M (2016) A comparison of measurement accuracy for DAG, tcpdump and windump. ResearchGate

13. Baker F, Fairhurst G (2015) IETF recommendations regarding active queue management. RFC 7567. https://doi.org/10.17487/RFC7567. https://www.rfc-editor.org/info/rfc7567
14. Benáček P, Puš V, Kubátová H, Čejka T (2018) P-4-To-VHDL: automatic generation of high-speed input and output network blocks. Microprocess. Microsyst. 56:22–33
15. Berde P, Gerola M, Hart J, Higuchi J, Kobayashi M, Koide T, Lantz B, O'Connor B, Radoslavov P, Snow W, et al (2014) ONOS: towards an open, distributed SDN OS. In: Proceedings of the third workshop on Hot topics in software defined networking, pp 1–6
16. Betker A, Gamrath I, Kosiankowski D, Lange C, Lehmann H, Pfeuffer F, Simon F, Werner A (2014) Comprehensive topology and traffic model of a nationwide telecommunication network. J Opt Commun Netw 6(11):1038–1047
17. Roberto B, Thomas D, Felipe H, Mohamed A, Joao M, Saverio N, Hans-Joerg K (2013) Rethinking access networks with high performance virtual software brases. In: Second European workshop on software defined networks. IEEE, pp 7–12
18. Bishop J, Chareau J-M, Bonavitacola F (2018) Implementing 5G NR features in FPGA. In: 2018 European conference on networks and communications (EuCNC). IEEE, pp 373–377
19. Blendin J, Bendfeldt F, Poese I, Koldehofe B, Hohlfeld O (2018) Dissecting Apple's meta-CDN during an iOS update. In: Proceedings of the internet measurement conference. ACM, pp 408–414
20. Borromeo JC, Kondepu K, Andriolli N, Valcarenghi L (2020) An overview of hardware acceleration techniques for 5G functions. In: 2020 22nd international conference on transparent optical networks (ICTON). IEEE, pp 1–4
21. Bosshart P, Daly D, Gibb G, Izzard M, McKeown N, Rexford J, Schlesinger C, Talayco D, Vahdat A, Varghese G, Walker D (2014) P4: programming protocol-independent packet processors. ACM SIGCOMM Comput Commun Rev 44(3):87–95
22. Bosshart P, Gibb G, Kim H-S, Varghese G, McKeown N, Izzard M, Mujica F, Horowitz M (2013) Forwarding metamorphosis: fast programmable match-action processing in hardware for SDN. ACM SIGCOMM Comput Commun Rev 43(4):99–110
23. Budiu M, Dodd C (2017) The p416 programming language. ACM SIGOPS Oper Syst Rev 51(1):5–14
24. Carrel D, Evarts J, Lidl K, Mamakos LA, Simone D, Wheeler R (1999) A method for transmitting PPP over Ethernet (PPPoE). RFC 2516. https://www.rfc-editor.org/info/rfc2516
25. Chiosi M, Clarke D, Willis P, Reid A, et al (2012) Network functions virtualisation: an introduction, benefits, enablers, challenges & call for actions. In: SDN and OpenFlow world congress, pp 1–16
26. Cisco Systems, Inc. (2022) TRex - realistic traffic generator. https://trex-tgn.ciscocom/. Accessed 10 Feb 2022
27. Clark DD, Minshall G, Zhang L, Peterson L, Ramakrishnan KK, Wroclawski JT, Shenker S, Partridge C, Crowcroft J, Braden RT, Deering SE, Floyd S, Davie BS, Jacobson V, Estrin D (1998) Recommendations on queue management and congestion avoidance in the internet. RFC 2309. https://doi.org/10.17487/RFC2309. https://www.rfc-editor.org/info/rfc2309
28. Cohen A, Shrum Ed, Anschutz T (2011) Migration to ethernet-based broadband aggregation. Technical report TR-101. Broadband Forum
29. Cooper RB (1981) Queueing theory. In: Proceedings of the ACM'81 conference. ACM, New York, NY, USA, pp 119–122
30. NVIDIA Corporation (2021) NVIDIA BLUEFIELD-2 DPU. https://www.nvidia.com/content/dam/en-zz/Solutions/Data-Center/documents/datasheetnvidia-bluefield-2-dpu.pdf. Product brief
31. Adam C, Glenn G, Lockwood JW, Nick McK (2009) A packet generator on the NetFPGA platform. In: 17th IEEE symposium on field programmable custom computing machines. IEEE, pp 235–238
32. Cui A, Hertoghs Y (2012) Multi-service broadband network functional modules and architecture. Technical repoer TR-145. https://www.broadband-forum.org/technical/download/TR-145.pdf. Broadband Forum

33. De Schepper K, Bondarenko O, Tsang I-J, Briscoe B (2016) Pi2: a linearized aqm for both classic and scalable tcp. In: Proceedings of the 12th international on conference on emerging networking experiments and technologies, pp 105–119
34. Dietz T, Bifulco R, Manco F, Martins J, Kolbe H-J, Huici F (2015) Enhancing the BRAS through virtualization. In: Proceedings of the 2015 1st IEEE conference on network softwarization (NETSOFT). IEEE, pp 1–5
35. Duckett M, Moisand J, Anschutz T, Kourkouzelis D, Arberg P (2006) Accommodating a maximum transit unit/maximum receive unit (MTU/MRU) greater than 1492 in the point-to-point protocol over ethernet (PPPoE). RFC 4638. https://doi.org/10.17487/RFC4638. https://www.rfceditor.org/info/rfc4638
36. Emmerich P, Gallenmüller S, Raumer D, Wohlfart F, Carle G (2015) MoonGen: a scriptable high-speed packet generator. In: Proceedings of the internet measurement conference. IMC '15. ACM, New York, NY, USA, pp 275–287
37. Emmerich P, Raumer D, Gallenmüller S, Wohlfart F, Carle G (2018) Throughput and latency of virtual switching with open vswitch: a quantitative analysis. J. Netw. Syst. Manag 26(2):314–338
38. Feamster N, Rexford J, Zegura E (2014) The road to SDN: an intellectual history of programmable networks. ACM SIGCOMM Comput Commun Rev 44(2):87–98
39. Fidler M, Rizk A (2014) A guide to the stochastic network calculus. IEEE Commun Surv & Tutor 17(1):92–105
40. Open Networking Foundation (2022) Software-Defined Networking (SDN) definition. https://opennetworking.org/sdn-definition/. Accessed 29 Jan 2022
41. The Linux Foundation (2010) https://www.dpdk.org/
42. Galles M, Matus F (2021) Pensando distributed services architecture. IEEE Micro 41(2):43–49
43. Garcia-Saavedra A, Costa-Pèrez X (2021) O-RAN: disrupting the virtualized RAN ecosystem. IEEE Commun Stand Mag 5(4):96–103
44. Gärtner C, Rizk A, Koldehofe B, Hark R, Guillaume R, Kundel R, Steinmetz R (2021) POSTER: leveraging PIFO queues for scheduling in time-sensitive networks. In: Proceedings of the international symposium on local and metropolitan area networks (LANMAN). IEEE, pp 1–2
45. Christoph G, Amr R, Boris K, Rhaban H, René G, Ralf S (2021) Leveraging flexibility of time-sensitive Networks for dynamic Reconfigurability. In: IFIP networking conference (IFIP Networking). IEEE, pp 1–6
46. Grazia CA, Patriciello N, Klapez M, Casoni M (2017) A cross-comparison between TCP and AQM algorithms: which is the best couple for congestion control. In: 2017 14th international conference on telecommunications (ConTEL), pp 75–82
47. Khronos Group (2021) OpenCL 2.2 Reference Pages. https://www.khronos.org/registry/OpenCL/sdk/2.2/docs/man/html/. Accessed 07 Feb 2022
48. gRPC: a high performance open source universal RPC framework (2015) https://grp.cio/. Accessed 07 Feb 2022
49. Guo C, Wu H, Deng Z, Soni G, Ye J, Padhye J, Lipshteyn M (2016) RDMA over commodity ethernet at scale. In: Proceedings of the 2016 ACM SIGCOMM conference, pp 202–215
50. Haleplidis E, Pentikousis K, Denazis S, Salim JH, Meyer D, Koufopavlou O (2015) Software-Defined Networking (SDN): layers and architecture terminology. RFC 7426. https://www.rfc-editor.org/info/rfc7426
51. Hasanin H, Mu J, Priest HR, Wolfgang K (2019) Towards understanding the performance of p4 programmable hardware. In: ACM/IEEE symposium on architectures for networking and communications systems (ANCS). IEEE, pp 1–6
52. Hauser F, Häberle M, Merling D, Lindner S, Gurevich V, Zeiger F, Frank R, Menth M (2021) A survey on data plane programming with P4: fundamentals, advances, and applied research. arXiv:2101.10632
53. Hawilo H, Shami A, Mirahmadi M, Asal R (2014) NFV: state of the art, challenges, and implementation in next generation mobile networks (vEPC). IEEE Netw 28(6):18–26

54. Huang G, Hu S, Qin F (2019) Yang data model for configuration interface of control-plane and user-plane separation BNG. https://toolsietf.org/id/draft-cuspdt-rtgwg-cu-separation-yang-model-04.html
55. Ibanez S, Brebner G, McKeown N, Zilberman N (2019) The p4-> netfpga workflow for line-rate packet processing. In: Proceedings of the 2019 ACM/SIGDA international symposium on field-programmable gate arrays. 2019, pp 1–9
56. IEEE Standard for Verilog Hardware Description Language (2006) In: IEEE Std 1364-2005 (Revision of IEEE Std 1364-2001), pp 1–590
57. IEEE Standard for VHDL Language Reference Manual (2019) In: IEEE Std 1076- 2019, pp 1–673
58. Intel, Inc. (2022) Intel Tofino 2. https://www.intel.de/content/www/de/de/products/network-io/programmable-ethernet-switch/tofino-2-series.html. Accessed 06 Feb 2022
59. Jain R (2004) Quality of experience. IEEE Multimed 11(1):95–96
60. Jiang H, Wang Y, Lee K, Rhee I (2012) Tackling bufferbloat in 3G/4G networks. In: Proceedings of the internet measurement conference. IMC '12. ACM, New York, NY, USA, pp 329–342
61. John W, Kern A, Kind M, Skoldstrom P, Staessens D, Woesner H (2014) SplitArchitecture: SDN for the carrier domain. IEEE Commun Mag 52(10):146–152
62. Juniper (2016) Warp17: the stateful traffic generator. https://github.com/Juniper/warp17
63. Kalia A, Zhou D, Kaminsky M, Andersen DG (2015) Raising the bar for using GPUs in software packet processing. In: 12th USENIX symposium on networked systems design and implementation (NSDI 15). USENIX Association, Oakland, CA, pp 409–423
64. Kendall DG (1953) Stochastic processes occurring in the theory of queues and their analysis by the method of the imbedded Markov chain. Ann Math Stat 338–354
65. Kerpez KJ, Cioffi JM, Ginis G, Goldburg M, Galli S, Silverman P (2014) Software-defined access networks. IEEE Commun Mag 52(9):152–159
66. Kfoury EF, Crichigno J, Bou-Harb E (2021) An exhaustive survey on P4 programmable data plane switches: taxonomy, applications, challenges, and future trends. In: IEEE Access
67. Khalil-Hani M, Nambiar VP, Marsono MN (2010) Hardware acceleration of OpenSSL cryptographic functions for high-performance internet security. In: 2010 international conference on intelligent systems, modelling and simulation. IEEE, pp 374–379
68. Khoshnevisan M, Joseph V, Gupta P, Meshkati F, Prakash R, Tinnakornsrisuphap P (2019) 5G industrial networks with CoMP for URLLC and time sensitive network architecture. IEEE J Selected Areas Commun 37(4):947–959
69. Kleinrock L (1975) Queueing systems, vol 1: Theory. Wiley-Interscience
70. Kreutz D, Ramos FMV, Verissimo PE, Rothenberg CE, Azodolmolky S, Uhlig S (2014) Software-defined networking: a comprehensive survey. In: IEEE, pp 14–76
71. Kundel R, Nobach L, Blendin J, Maas W, Zimber A, Kolbe H-J, Schyguda G, Gurevich V, Hark R, Koldehofe B, Steinmetz R (2021) OpenBNG: central office network functions on programmable data plane hardware. Int J Netw Manag 31(1):1–25
72. Kundel R, Rizk A, Blendin J, Koldehofe B, Hark R, Steinmetz R (2021) P4-CoDel: experiences on programmable data plane hardware. In: Proceedings of the IEEE international conference on communications (ICC). IEEE, pp 1–6
73. Kundel R, Wallerich J, Maas W, Nobach L, Koldehofe B, Steinmetz R (2019) Queueing at the telco service edge: requirements, challenges and opportunities. In: Proceedings of the 1st workshop on buffer sizing. Stanford University, pp 1–6
74. Lam CF (2011) Passive optical networks: principles and practice. Elsevier
75. Langley A, Riddoch A, Wilk A, Vicente A, Krasic C, Zhang D, Yang F, Kouranov F, Swett I, Iyengar J, et al (2017) The QUIC transport protocol: design and internet-scale deployment. In: Proceedings of the conference of the ACM special interest group on data communication, pp 183–196
76. Boudec J-YL, Thiran P (2001) Network calculus: a theory of deterministic queuing systems for the internet. Springer

77. Zexian L, Uusitalo MA, Shariatmadari H, Singh B (2018) 5G URLLC: design challenges and system concepts. In: 15th international symposium on wireless communication systems (ISWCS). IEEE, pp 1–6
78. Maeder A, Lalam M, De Domenico A, Pateromichelakis E, Wübben D, Bartelt J, Fritzsche R, Rost P (2014) Towards a flexible functional split for cloud-RAN networks. In: 2014 European conference on networks and communications (EuCNC). IEEE, pp 1–5
79. Makinen L, Nurminen JK (2008) Measurements on the feasibility of TCP NAT traversal in cellular networks. In: Next generation internet networks. IEEE, pp 261–267
80. Marchese Mario (2007) QoS over heterogeneous networks. Wiley
81. McKeown N, Anderson T, Balakrishnan H, Parulkar G, Peterson L, Rexford J, Shenker S, Turner J (2008) OpenFlow: enabling innovation in campus networks. ACM SIGCOMM Comput Commun Rev 38(2):69–74
82. Metz C (2022) Mavericks invent future internet where Cisco is meaningless. https://www.wired.com/2012/04/nicira/. Accessed 29 Jan 2022
83. Micheel J, Donnelly S, Graham I (2001) Precision timestamping of network packets. In: Proceedings of the 1st ACM SIGCOMM workshop on internet measurement. IMW '01. ACM, pp 273–277
84. Morton A (2017) Considerations for benchmarking virtual network functions and their infrastructure. RFC 8172. https://doi.org/10.17487/RFC8172. https://www.rfc-editor.org/info/rfc8172
85. Nagy B, Orosz P, Tóthfalusi T, Kovács L, Varga P (2018) Detecting DDoS attacks within milliseconds by using FPGA-based hardware acceleration. In: IEEE/IFIP network operations and management symposium. IEEE, pp 1–4
86. Neshatpour K, Malik M, Ghodrat MA, Sasan A, Homayoun H (2015) Energy-efficient acceleration of big data analytics applications using FPGAs. In: 2015 IEEE international conference on big data. IEEE, pp 115–123
87. Netronome Systems, Inc. (2018) Netronome 25GbE SmartNICs with open vSwitch hardware offload. White Paper. https://www.netronome.com/media/documents/WP_Netronome_25GbE_SmartNICs_with_Open_vSwitch_Hardware_Offload.pdf
88. Netronome Systems, Inc. (2018) P4 Data plane programming for server-based networking applications. White Paper. https://www.netronome.com/media/documents/WP_P4_Data_Plane_Programming.pdf
89. Nichols K, Jacobson V, McGregor A, Iyengar J (2018) Controlled delay active queue management. RFC 8289. https://doi.org/10.17487/RFC8289. https://www.rfc-editor.org/info/rfc8289
90. Nielsen J (2022) Xilinx vitis unified software platform. https://www.xilinx.com/products/design-tools/vitis/vitis-platform.html. Accessed 28 Jan 2022
91. NVIDIA Corporation (2022) CUDA toolkit documentation v11.6.0. https://docs.nvidia.com/cuda/. Accessed 07 Feb 2022
92. P4Runtime Specification version 1.3.0 (2021) https://p4.org/p4-spec/p4runtime/main/P4Runtime-Spec.html. Accessed 07 Feb 2022
93. Pan R, Natarajan P, Baker F, White G (2017) Proportional integral controller enhanced (PIE): a lightweight control scheme to address the bufferbloat problem. RFC 8033. https://doi.org/10.17487/RFC8033. https://www.rfceditor.org/info/rfc8033
94. Papagiannaki K, Moon S, Fraleigh C, Thiran P, Diot C (2003) Measurement and analysis of single-hop delay on an IP backbone network. IEEE J Selected Areas Commun 21(6):908–921
95. Paxson V, Allman M, Stevens WR (1999) TCP congestion control. RFC 2581. https://doi.org/10.17487/RFC2581. https://www.rfceditor.org/info/rfc2581
96. Pensando, Inc. (2021) Pensando DSC-25; Distributed services card. https://pensando.io/wp-content/uploads/2020/03/DSC-25-Product-Brief-v05.pdf. Product brief
97. Peterson L, Al-Shabibi A, Anshutz T, Baker S, Bavier A, Das S, Hart J, Palukar G, Snow W (2016) Central office rearchitected as a data center. IEEE Commun Mag 54(10):96–101
98. Pfaff B, Pettit J, Koponen T, Jackson E, Zhou A, Rajahalme J, Gross J, Wang A, Stringer J, Shelar P, et al (2015) The design and implementation of open vswitch. In: 12th USENIX symposium on networked systems design and implementation (NSDI 15), pp 117–130

99. Pfrommer J, Ebner A, Ravikumar S, Karunakaran B (2018) Open source OPC UA PubSub over TSN for realtime industrial communication. In: IEEE 23rd international conference on emerging technologies and factory automation (ETFA), vol 1. IEEE, pp 1087–1090
100. Pham M, Hoang DB (2016) SDN applications - the intent-based Northbound Interface realisation for extended applications. In: 2016 IEEE NetSoft conference and workshops (NetSoft), pp 372–377
101. Plexxi, Inc. (2016) Latency in ethernet switches. White Paper
102. Primorac M, Bugnion E, Argyraki K (2017) How to measure the killer microsecond. In: Proceedings of the workshop on kernel-bypass networks. KBNets '17. ACM, pp 37–42
103. Ranaweera C, Wong E, Nirmalathas A, Jayasundara C, Lim C (2017) 5G C-RAN with optical fronthaul: an analysis from a deployment perspective. J Lightwave Technol 36(11):2059–2068
104. Ricart-Sanchez R, Malagon P, Salva-Garcia P, Perez EC, Wang Q, Calero JMA (2018) Towards an FPGA-accelerated programmable data path for edge-to-core communications in 5G networks. J Netw Comput Appl 124:80–93
105. Rodriguez D (2018) Next generation central offices transform network edge with datacenter economics, cloud flexibility. Blog post. Accessed 03 Feb 2022. https://builders.intel.com/blog/next-generation-central-office-transform-network-edge/
106. Rumble SM, Ongaro DD, Stutsman R, Rosenblum M, Ousterhout JK (2011) It's time for low latency. In: 13th workshop on hot topics in operating systems (HotOS XIII). USENIX Association
107. Schmitt JB (2001) Heterogeneous network quality of service systems. Springer Science & Business Media
108. Schwarzkopf F, Veith S, Menth M (2016) Performance analysis of CoDel and PIE for saturated TCP sources. In: 28th international teletraffic congress (ITC 28), vol 01. IEEE, pp 175–183
109. Sharma NKr, Zhao C, Liu M, Kannan PG, Kim C, Krishnamurthy A, Sivaraman A. Programmable calendar queues for high-speed packet scheduling. In: 17th USENIX symposium on networked systems design and implementation (NSDI 20). USENIX Association, Santa Clara, CA, pp 685–699
110. Shrivastav V (2019) Fast, scalable, and programmable packet scheduler in hardware. In: Proceedings of the ACM special interest group on data communication, pp 367–379
111. Shrum E, Allan D, Thorne D (2004) Broadband Remote Access Server (BRAS) requirements document. Technical report TR-092. Broadband Forum
112. Singh SK, Rothenberg CE, Patra G, Pongracz G (2019) Offloading virtual evolved packet gateway user plane functions to a programmable ASIC. In: Proceedings of the 1st ACM CoNEXT workshop on emerging In-Network computing paradigms. ENCP '19. ACM, pp 9–14
113. Sivaraman A, Subramanian S, Alizadeh M, Chole S, Chuang S-T, Agrawal A, Balakrishnan H, Edsall T, Katti S, McKeown N (2016) Programmable packet scheduling at line rate. In: Proceedings of the 2016 ACM SIGCOMM conference, pp 44–57
114. Sivaraman A, Winstein K, Subramanian S, Balakrishnan H (2013) No silver bullet: extending SDN to the data plane. In: Proceedings of the twelfth ACM workshop on hot topics in networks, pp 1–7
115. Sourouri M, Gillberg T, Baden SB, Cai X (2014) Effective multi-GPU communication using multiple CUDA streams and threads. In: 2014 20th IEEE international conference on parallel and distributed systems (ICPADS). IEEE, pp 981–986
116. Spirent Communications, Inc (2012). Spirent testcenter virtual. Product brief
117. Steinmetz R, Nahrstedt K (2004) Multimedia systems. Springer Science & Business Media
118. Sun W, Ricci R (2013) Fast and flexible: parallel packet processing with GPUs and click. In: Architectures for networking and communications systems, pp 25–35
119. Sundaresan S, De Donato W, Feamster N, Teixeira R, Crawford S, Pescapè A (2011) Broadband internet performance: a view from the gateway. ACM SIGCOMM Comput Commun Rev 41(4):134–145
120. Tan L, Wei S, Zhang W, Lv J, Zhang Z, Miao J, Liu X, Li N (2021) In-band network telemetry: a survey. Comput Netw 186:1–24

121. Tanenbaum AS (2011) Computer networks, 5th edn. Pearson Education India
122. Timmers M, Guenach M, Nuzman C, Maes J (2013) G. fast: evolving the copper access network. IEEE Commun Mag 51(8):74–79
123. Touger E (2018) What is an FPGA and why is it a big deal? https://www.prowesscorp.com/what-is-fpga/. Accessed 06 Feb 2022
124. Vasiliadis G, Koromilas L, Polychronakis M, Ioannidis S (2014) GASPP: a GPU-accelerated stateful packet processing framework. In: Gibson G, Zeldovich N (ed) 2014 USENIX annual technical conference ATC. USENIX Association, pp 321–332
125. Vrdoljak M, Vrdoljak SI, Skugor G (2000) Fixed-mobile convergence strategy: technologies and market opportunities. IEEE Commun Mag 38(2):116–121
126. Wang F-Y, Yang L, Cheng X, Han S, Yang J (2016) Network softwarization and parallel networks: beyond software-defined networks. IEEE Netw 30(4):60–65
127. Wang H, Soulè R, Dang HT, Lee KS, Shrivastav V, Foster N, Weatherspoon H (2017) P4fpga: a rapid prototyping framework for p4. In: Proceedings of the symposium on SDN research, pp 122–135
128. Watfa MK, Diab M, Stephen N (2015) Improving TCP performance in mix networks. In: Progress in systems engineering. Springer, pp 423–428
129. Watson G, McKeown N, Casado M (2006) NetFPGA: a tool for network research and education. In: 2nd workshop on architectural research using FPGA platforms (WARFP), vol 3
130. Xilinx, Inc. (2019) Xilinx Alveo powers OVS OFFLOAD. Solution brief. https://www.xilinx.com/publications/solution-briefs/partner/vvdnovs-solution-brief.pdf
131. Zerwas J, Kellerer W, Blenk A (2021) What you need to know about optical circuit reconfigurations in datacenter networks. In: 2021 33th international teletraffic congress (ITC-33), pp 1–9
132. Zhang C, Chen Z, Song H, Yao R, Xu Y, Wang Y, Miao J, Liu B (2021) PIPO: efficient programmable scheduling for time sensitive networking. In: 2021 IEEE 29th international conference on network protocols (ICNP). IEEE, pp 1–11
133. Zhang Q, Liu V, Zeng H, Krishnamurthy A (2017) Highresolution measurement of data center microbursts. In: Proceedings of the internet measurement conference. ACM, pp 78–85

Chapter 3
Design of QoS-Aware Network Functions

In this chapter, we focus on realizing network functions with programmable hardware to achieve a high Quality of Service (QoS) in Internet access networks and in general.

First, in Sect. 3.1, we introduce the *host bypassing* approach, improving the performance and flexibility of PCIe-based hardware accelerator cards, i.e., Field Programmable Gate Arrays (FPGAs) and Graphics Processing Units (GPUs), for network function acceleration. Second, in Sect. 3.2, we investigate hardware acceleration capabilities in Internet access networks, specifically considering the User Plane Function (UPF) in 5G networks and the Broadband Network Gateway (BNG) in residential access networks. Last, the realization of Active Queue Management (AQM) algorithms in hardware is challenging but necessary to provide a high QoS in Internet access networks. Therefore, we examine these challenges in Sect. 3.3 and deduce general guidelines on the example of the Controlled Delay (CoDel) AQM algorithm.

3.1 Host Bypassing: PCI Express Hardware Accelerator Integration

One main challenge of PCIe-based hardware accelerators for network functions is the input and output of data [20]. Peripheral Component Interconnect Express (PCIe) is the de-facto standard for integrating peripheral components in computer systems, including hardware accelerators [38]. While at least some FPGA accelerators offer Ethernet connectors in addition to PCIe, GPUs provide either only display connectivity or no additional connector to PCIe at all. Therefore, data to be processed must be injected via PCIe.

R. Kundel, *Accelerating Network Functions Using Reconfigurable Hardware*,
Springer Theses, https://doi.org/10.1007/978-3-031-52872-9_3

3.1.1 General Host Bypassing Concept

Figure 3.1 depicts the general architecture of hardware-accelerated network functions in commodity servers. Within the server, a Network Interface Card (NIC) and a *hardware accelerator* are connected as a PCIe endpoint to the PCIe infrastructure of the server via one or multiple lanes. Typically, these PCIe lanes are directly connected to the CPU; however, more complex bus-topologies can exist, e.g., the utilization of a PCIe switch that increases the total number of connectable devices. All PCIe endpoints and the system main memory share the same *physical memory address space*. A software thread running on one of the CPU cores can access the PCIe endpoint device's memory by reading or writing on a physical memory address. Additionally, PCIe endpoints can perform Direct Memory Access (DMA) reads and writes within this physical memory address space.

The general process for packet I/O in state-of-the-art approaches with hardware-accelerated network functions is as follows: (1) A packet enters on one of the Ethernet ports of the NIC and is written via DMA into the system's main memory. (2) From there, a software process, running on one of the CPU cores, copies the packet into the hardware accelerator or initiates a DMA transfer. Now, the packet is within the hardware accelerator and can be processed by the actual network function, denoted as f_x. (3) After being processed, the packet to be sent will be transferred back into the main memory. (4) Last, a software process initiates the transmission of the packet, and the packet is transferred via PCIe to the NIC.

In total, the system copies the packet four times over the PCIe bus, which causes significant overhead on bus resources and CPU utilization without any additional value. To that end, we present the *host bypassing* approach as shown in Fig. 3.1. Network packets are transferred directly from the NIC into the hardware accelerator, without any interaction of the system's main memory and one or more Central Processing Unit (CPU) threads. In general, this is called a PCIe *peer-to-peer* transfer, allowing one endpoint to read or write data directly from or into another endpoint [4, 42]. By that, the total number of memory copies is reduced from four to two. As the main memory of a system is shared between all applications, utilizing it as temporary transfer storage it can quickly exceed its maximum memory bandwidth and conse-

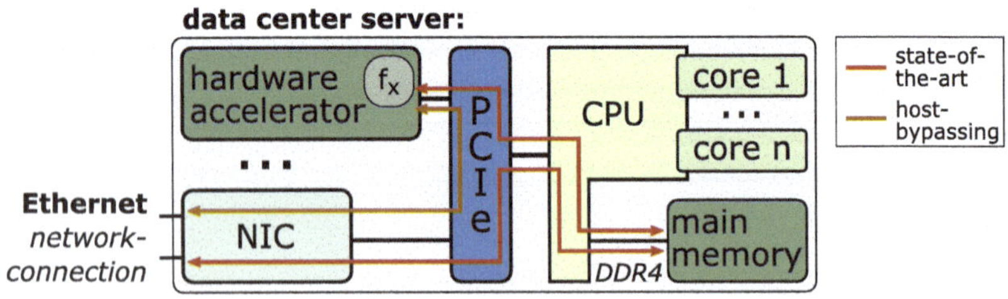

Fig. 3.1 Server with PCIe-based hardware accelerator. The yellow line indicates the host bypassing data path, not involving the host memory. ©2021 IEEE. Reprinted, with permission, from "Host Bypassing: Direct Data Piping from the Network to the Hardware Accelerator" [21]

quently limit other applications running on the CPU only. Further, main memory is typically realized with Dynamic Random Access Memory (DRAM) memory technology, e.g., DDR4, having a non-deterministic behavior due to physical limitations of this memory kind [28]. This non-determinism can cause an increased latency jitter, e.g., caused by concurrent DRAM refresh cycles or simultaneous accesses from multiple processes on different memory regions.

As a consequence, the *host bypassing* approach is supposed to reduce the packet I/O latency and jitter while increasing the maximum achievable throughput. At the same time, commodity hardware accelerators not specially built for networking purposes, e.g., GPUs, can be integrated into the network data path. In contrast to a fixed-wired accelerator, this approach allows a flexible orchestration within a server by configuring the NIC and accelerator with the appropriate physical addresses of each other. In this work, we show the viability of this approach on the example of GPUs (Sect. 3.1.4) and FPGAs (Sect. 3.1.5). Still it can be generalized to any other hardware accelerator technology providing a DMA-capable PCIe interface.

3.1.2 Working Principle of Poll Mode Drivers

Design goals of *host bypassing* are flexibility and the utilization of commodity hardware, especially off-the-shelf NICs should be used. The advantage of user space poll-mode drivers over conventional NIC drivers is that the host operating system is not involved in any packet I/O; the user space process communicates directly with the NIC. Therefore, we reverse-engineered the behavior of an existing poll-mode driver, being open-source available as part of the DPDK framework [10]. Specifically, we choose the *Intel 82599* chipset in this work, which is currently one of the most widespread NICs in data centers. This NIC is controlled by the `ixgbe` DPDK-driver.

Figure 3.2 shows the concept of poll-mode drivers, including DPDK. The NIC and the user space application communicate over a shared memory region in the system's main memory. This memory region is pinned to a fixed physical memory address, pointing on a not-fragmented memory huge page. If the physical memory gets re-fragmented by the operating system, the user space application, operating on virtual addresses, would not be affected by this. However, the NIC writes on the old physical address, and therefore errors will occur. Thus, this memory must be pinned to a fixed physical address.

Within the shared memory, the received packets and packets to be sent are handed over between the NIC and the user space application. The packets are stored in a shared memory buffer, labeled *mbuf*. In addition, multiple descriptor rings of constant size for receiving and sending exist, at least one for each. Each ring has two pointers, a head and tail pointer. The ring sector between the tail and head pointer (in a clockwise direction) is controlled by the user space application and the remaining part by the NIC. Thus, the NIC and the *application* can work without any interference on the same rings.

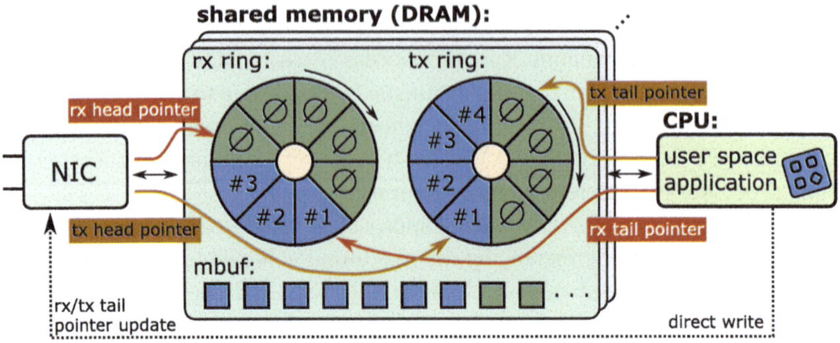

Fig. 3.2 Memory mapped interaction between user space network function and DPDK-capable NIC. ©2021 IEEE. Reprinted, with permission, from "Host Bypassing: Direct Data Piping from the Network to the Hardware Accelerator" [21]

63 … 1	0
Packet Buffer Address [63:1]	A0
Header Buffer Address [63:1]	DD

Fig. 3.3 Rx descriptor format for Intel 82599 NICs according to: [17]. The A0 bit is unused, the DD bit (descriptor done) is used to mark a processed descriptor

The process of receiving packets works as follows:

(1) The user space application allocates a memory block with at least the Maximum Transmission Unit (MTU) size in the *mbuf* and writes its address into a descriptor entry, formatted as shown in Fig. 3.3. Note that the Descriptor Done (DD) bit must be set to zero. Last, the application increases the *rx tail pointer* in the NIC to indicate the change.

(2) The NIC reads the memory address from the descriptor ring at the *rx head pointer* and writes the next arriving packet on the provided memory address. To notify the application, the NIC overwrites the descriptor ring entry with metadata information of the received packet, e.g., the packet length, and sets the *DD* bit to one.

(3) The receiving thread of the user space polls the *DD* bit of the descriptor ring entry of the tail pointer to await the next packet. This process is named "busy waiting", as a single CPU core is fully utilized by permanently checking the status. Therefore, poll-mode drivers are only reasonable for high loads when the overall efficiency is higher than for an interrupt solution. As soon as the *DD*-bit flips to one, the receiving thread can read the packet from the memory buffer or forward a pointer of it for further processing. For this, the memory address must be saved at allocation time, as the NIC has overwritten the descriptor ring entry.

(4) Last, the software process prepares the descriptor for a new receiving packet and increases the rx tail pointer as described in step (1).

```
pointer rx_descriptor_ring [256]
pointer packet_bufs [256]
receive_thread():
    //Initialize empty descriptors:
    for i in range (0,256):
        rx_descriptor_ring[i].dd = 0
        pointer tmp = malloc(MTU)
        rx_descriptor_ring[i].pkt_buf_addr = tmp
        packet_bufs[i] = tmp
    int tailpointer = 0
    init_NIC(rx_descriptor_ring)
    while(true):
        //Polling for new packets:
        if(rx_descriptor_ring[tailpointer].dd == 1):
            network_function.process(packet_bufs[tailpointer])
            //Prepare the descriptor for receiving a new packet:
            pointer tmp = malloc(MTU)
            rx_descriptor_ring[i].pkt_buf_addr = tmp
            packet_bufs[tailpointer] = tmp
            //Increase tailpointer
            tailpointer++
            tailpointer = tailpointer % 256 //ring size
            update_NIC_tailpointer(tailpointer)
```

Listing 3.1 Simplified pseudo-code of a DPDK-application receiving network packets from the NIC via a descriptor ring with 256 entries located in shared memory

This receiving process is depicted as pseudo-code in Listing 3.1. In the example given in Fig. 3.2, currently, three packets are received by the NIC but not yet processed by the application. The remaining five slots are free for receiving future packets. In the case of the used NIC, the number of descriptors is higher and can be configured before runtime, i.e., between 64 and 256, and by that, packets can be processed in bulks. The descriptor of the used NIC allows specifying a second memory address to store the packet headers, separated from the payload of the packet. However, this work focuses on packet processing, including all headers, and therefore we do not enable this functionality.

3.1.3 Software Driver Modifications

The prototypes presented in this work build upon the DPDK framework, which required minor changes to initialize the NIC for *host bypassing*. For this, we added an Application Programming Interface (API) function to set custom rx and tx ring addresses in the NIC as well as the number of rings. With this functionality, the NIC can be reconfigured to read and write descriptor rings located anywhere in the physical memory address space of the system; in the approach of this work, they are located in the PCIe hardware accelerator.

Further, the hardware accelerator, i.e., the FPGA or GPU, must be configured similarly. First, the physical base address of the NIC must be set. Second, the hardware accelerator must know its own physical address to independently determine addresses in descriptor ring entries. Third, the initialization driver notifies the hardware accelerator that the NIC is ready, and tail pointer updates can be written.

In addition, the PCIe endpoints of the NIC and *hardware accelerator* must be allowed to communicate with each other. For this, it is required to load a kernel module for both of them, allowing to perform DMA bus transfers. In our experiments, we utilized the general-purpose module *igb_uio*. Further, depending on the PCIe architecture of the server, it might be required to disable PCIe access control services.

These initialization procedures must be performed only once before starting the system. At runtime, no interaction by any control application running on the CPU is needed, and therefore this resource can be utilized with other tasks.

The driver modifications, build instructions, and the source code for the FPGA- and GPU-based prototype are publicly available on Github [15]. In the following two subsections, we will introduce the details on the realization of *host bypassing* on FPGAs and GPUs.

3.1.4 FPGA-Based Host Bypassing

FPGAs differ significantly from their internal architecture and way of solving a given problem from conventional computer processors as introduced in Sect. 2.3. Therefore, existing poll-mode drivers for NICs can not be migrated in a straightforward manner on FPGAs to enable *host bypassing*, i.e., by cross-compiling from one language to another. Instead, a hardware design of logical circuits must be described, providing the same behavior to the NIC as a software-based driver. In this work, we built upon one of the most prominent hardware description languages, *Verilog*. Yet, other languages provide similar expressiveness and would probably not influence the results strongly as the described logic remains similar [16]. The presented hardware design does not contain any vendor-specific functionality and can be realized with FPGA boards from Xilinx and Intel, the two most prominent vendors in the world. We tested the presented design with multiple FPGA technologies, including the Xilinx Virtex 7, Xilinx Virtex UltraScale+, and Intel Stratix 10 GX chipsets. However, the concepts can be applied to any other platform providing PCIe connectivity and sufficient programmable logic circuits. The presented design is synthesized with a clock frequency of 250 MHz. In the following subsections, we introduce the key concepts, design decisions, and implementation details. The overall design is depicted in Fig. 3.4.

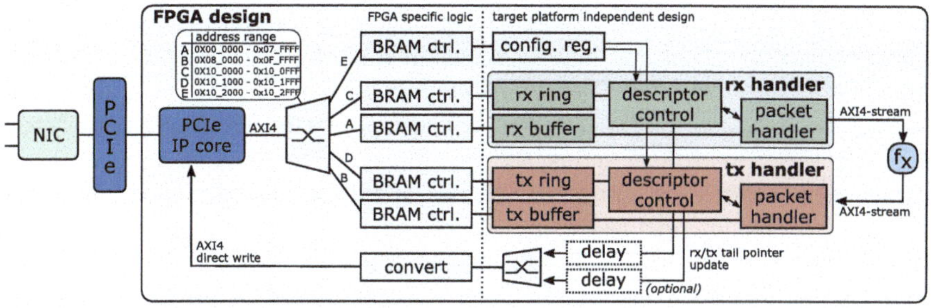

Fig. 3.4 Functional components within the FPGA, required for realizing the host bypassing functionality. The accelerated network function is abstracted as f_x. ©2021 IEEE. Reprinted, with permission, from "Host Bypassing: Direct Data Piping from the Network to the Hardware Accelerator" [21]

Shared Memory

The main principle of *host bypassing* with FPGAs is memory mapping. As described in Sect. 3.1.2, the NIC communicates with the software driver over a shared memory region, located in the system's main memory and is reachable over physical addressing on the PCIe bus. This shared memory region should be located inside the FPGA, still accessible via physical addressing on the PCIe bus for the NIC but also for logic within the FPGA.

In Fig. 3.4, we present the component overview within the FPGA. A PCIe interface is required to enable the aforementioned PCIe-accessible shared memory region. For this, one must build upon an Intellectual Property Core (IP core), as the PCIe logic is too complex and the timing requirements are too high for a custom FPGA design. However, most FPGA vendors provide this functionality as an IP core, realized in partially fixed logic within the FPGA, and therefore this is no limitation. The PCIe IP core offers a generic memory interface consisting of address, data wires for read and write operations, and various control signals, attached to any memory-like module. Assuming a data wire width of 64 bits and a FPGA clock frequency of 200 MHz, the theoretical maximum throughput of this interface can be determined by: $64\,\text{bit} \cdot 200 \cdot 10^6 \frac{1}{s} = 12.8\,\text{Gbit/s}$. This number must be higher than the required bandwidth as all network packets plus control overhead are transmitted over this interface. Thus, the PCIe IP core must be configured to a PCIe technology, e.g., *PCIe Gen3*, a sufficient number of lanes, e.g., $x8$, and an appropriate data bus width, e.g., 64 bits or 128 bits, fulfilling this bandwidth demand. At the startup time of the host system, the PCIe IP core requests a physical address range from the PCIe system, e.g., 2 MB are sufficient for *host bypassing* with up to 256 descriptor ring entries, but a larger address range is not detrimental.

Inside the FPGA, the memory interface of the IP core, e.g., the *AXI4* interface for Xilinx and the *Avalon* interface for Intel FPGAs, is mapped on multiple memory regions. For this, we utilize a crossbar implementation that forwards the incoming read and write requests based on the address to one of the five memory controllers.

The table in the top left of Fig. 3.4 depicts the mapping of physical addresses to the different controllers. Note that these addresses are relative to the physical start address of the FPGA. Assuming the FPGA address region starts at $0x0800\_0000$, a physical write access on the address $0x0800\_0042$ would appear at the FPGA internal data bus at the address $0x0000\_0042$ and is therefore forwarded to the *rx buffer*, denoted as A. Each memory region is abstracted by a very simple memory interface, consisting of an address bus, the read and write data bus, and read- and write-enable signals. This interface is very common to access Block Random Access Memory (BRAM) blocks and is therefore named BRAM interface. The *BRAM controller* converts the complex but universal *AXI4* signals to this primitive memory interface. This interface represents the cutting point between FPGA-type specific and FPGA-type generic hardware modules.

On the right side of the BRAM interface, four memories, namely the *rx ring*, the *rx buffer*, the *tx ring*, and the *tx buffer*, are realized with FGPA internal BRAM memory blocks, consisting of fast Static Random Access Memory (SRAM) cells. Each of these blocks can be mapped to the corresponding DPDK memory region as shown in Fig. 3.2. Note that we split up the *mbuf* memory into two buffer memories, one for sending and one for receiving packets, as the underlying processes are also totally autonomous from each other. These four memories are independent of each other and do not share any resources. Further, they provide each two BRAM interfaces, also called *dual-port RAM*, allowing isochronous access from the module attached on the left and on the right side without any synchronization overhead. Thus, the control logic of the rx and tx handler can independently from incoming PCIe accesses read and write from one of the four memories.

Further, a *configuration register* exists, which can be accessed via PCIe from the host. In this register, the software driver can write the physical address of the NIC and of the FPGA. For details on the configuration, we refer the reader to Sect. 3.1.3.

FPGA-Based Poll Mode Driver

The receiving and sending logic can be examined independently. In general, the behavior of the FPGA implementation reproduces the behavior of the reference software implementation as described in Sect. 3.1.2.

The *rx ring* must be first initialized with valid physical memory addresses within the *rx buffer* for receiving packets. For this, the physical address of the FPGA, known from the software driver initialization, is used as an offset. After all descriptor ring entries are initialized and the software driver configured the NIC with the GPU address, the NIC updates once the *rx tail pointer* to point on the first valid ring entry.

Now, the first packet can be received. The NIC reads one or multiple descriptor ring entries via DMA from the FPGA. As soon as a packet arrives at the Ethernet port of the NIC, it will be written via DMA on the physical memory address of the first descriptor ring entry, the current *rx head pointer* is pointing on. In contrast to the main memory from server systems, the FPGA has the opportunity to monitor incoming data bursts, allowing the detection of new arriving packets. This writing process can be distributed over one or multiple PCIe bus transfers. Therefore, the FPGA can not determine from incoming bytes that a new packet was received entirely. Consequently, analogous to software-based poll mode drivers, the FPGA must await the rx descriptor writeback. As soon as the NIC has written all data in the *rx buffer* of the FPGA, it updates the *rx ring* entry accordingly. This writeback information contains the byte length of the received packet and further metadata, including the *DD* bit, which indicates the successful reception of the packet (Compare Fig. 3.3). The *rx descriptor control* permanently polls this descriptor ring entry, awaiting the *DD* bit to be active. Due to the dual-port BRAM, the descriptor control can access this memory address every clock cycle without affecting the performance on the left side of the memory towards the PCIe interface.

Next, the *descriptor control* hands over the extracted packet length from the writeback descriptor and the memory address within the *rx buffer* to the *packet handler*. The *packet handler* reads the packet data from the given address in the *rx buffer* and provides this as a serialized stream, compliant with the *AXI4-stream* and *Avalon-stream* protocol specifications. This stream interface is the current de-facto standard packet processing in FPGAs; however, providing the data in any other format would not change the presented concept significantly. Now, the network function f_x can perform any packet processing, which is not the focus of this work. Last, the *descriptor control* must reset the descriptor ring entry with a new valid physical address. As the packets are never stored longer within the *rx buffer*, we map the descriptor ring entries one-to-one on memory slots in the buffer memory. This simplifies the memory allocation and memory address calculation strongly. The NIC is informed by writing the updated *rx tail pointer* via PCIe into a control register of the NIC. Updating the rx and tx tail pointers are the only PCIe operations initiated by the FPGA, and for this, the physical address of the NIC must be known. As the *rx ring* consists of many entries, this can be done only for every n_{th} packet or every m time units, reducing the total traffic on the PCIe bus. For this, an optional *delay* module for tail pointer updates is instantiated between the *descriptor control* and the PCIe IP core.

Analogous to receiving packets, the *tx handler* prepares packets to be transmitted by the NIC. Packets are provided on an *AXI4-stream* interface to the packet handler, which writes the data into the next free memory slot of the *tx buffer*. Identical to the *rx handler*, one memory slot in the buffer corresponds to one entry in the *tx ring*. Next, the packet length and location in the buffer are handed over to the *tx descriptor control*. From there, the packet address and length is written to the next free *tx ring* entry, indicated by the current *tx tail pointer* (Compare Fig. 3.3). Last, the updated *tx tail pointer* is written into the NIC to indicate the new packet to be sent. Similar to the receiving process, this pointer update can be delayed and grouped that not for every packet a PCIe transfer is required. However, in contrast to the receiving process, this causes additional latency as packets are sent later. Therefore, a timeout is mandatory to prevent packets from never being sent. For example, if the tail pointer is updated every 8_{th} packet, but only 7 packets are added to the *tx ring*, they will never be sent. A timer, starting when the first packet was added to the ring after a preceding *tail pointer* update, guarantees the packet to be sent within a timeout period, e.g., 1 μs.

Typically, the speed of sending packets by the NIC can be assumed to be higher than a software process can generate new data. In that case, the *tx ring* can not run over. However, a high-speed packet source, e.g., an FPGA performing *host bypassing*, can fill this ring faster than the packets can be sent. As a solution approach, an optional transmission confirmation of a packet can be enabled, realized by a final descriptor writeback from the NIC into the *tx ring*. For this functionality, a *writeback* bit must be set in the tx descriptor when handing over the packet to the NIC. By this, the *descriptor control* can ensure that a packet was sent before overwriting this ring entry with a new packet and is implemented as an optional feature in our concept prototype. However, as investigated later as part of the evaluation in Sect. 5.1.2, it is very likely that these additional PCIe transfers lower the total system performance.

Monitoring and Debugging

The main purpose of this prototype is to prove and understand the *host bypassing* concept. Therefore, investigating the internals of the design at runtime is crucial to understand performance implications and learn from them.

To enable this, we introduced the following three capabilities of on-chip monitoring: First, interfaces between the modules allow to monitor any interaction with the NIC by attaching an on-chip **logic analyzer**, provided as part of the synthesis toolchains by both used FPGA vendors.

Second, the standardized *AXI4-stream* interface allows connecting an **ethernet port** of the FPGA to receive packets via Ethernet and send them out via *host bypassing*. Analogous, the opposite direction can be leveraged. By this, the development processes is facilitated and benchmarking becomes more precise.

Third, the software driver, introduced in Sect. 3.1.3 for initialization, can also read on physical addresses at runtime and can be used to read out hardware registers for monitoring purposes, e.g., the *rx/tx head/tail pointers* to detect faulty behavior or lock states.

3.1.5 GPU-Based Host Bypassing

The approach, presented in the previous Sect. 3.1.4, realizes *host bypassing* with FPGAs and commodity NICs. With this, the packet I/O of FPGA-accelerated network functions can be improved compared to state-of-the-art FPGA acceleration approaches in PCIe environments. However, on the one hand, FPGAs are an accelerator technology that causes high development effort compared to conventional software development. Indeed, FPGAs can provide a good performance due to their reconfigurability on bit-level, and any behavior can be realized. On the other hand, FPGA accelerators are not as widespread as other acceleration technologies in today's data centers, e.g., GPUs. GPUs can execute a software program on many integrated and parallel processing cores, described by software programming languages, similar to those used in conventional software development. Due to their parallelism, they are very well suited for parallel packet processing.

Therefore, in the following of this section, we will investigate how the *host bypassing* approach can be extended from FPGAs to commodity GPUs. As GPUs' internals strongly differ from FPGAs, the previously introduced approach can not be applied one-to-one on GPUs; however, the main concept remains unchanged.

Memory Exposing on GPUs

Host bypassing with GPUs builds upon the same principle as with FPGAs: an exposed memory within the hardware accelerator, i.e., the GPU, accessible via DMA by the NIC. To enable this, we choose the *NVIDIA Quadro RTX 4000* GPU for our research, as most consumer cards currently do not allow exposing memory to be accessible by other PCIe devices. To access all features of this hardware, we select the vendor-specific programming language *CUDA* [36]. Note that this is not a technical limitation; instead, the vendor unlocked this feature only for professional GPUs, presumably for reasons of market separation [4].

To investigate the practicability of *host bypassing* on GPUs, we propose a prototypical design. The overall software architecture of this prototype is shown in Fig. 3.5, consisting of one big memory region in the global memory and three CUDA kernels, each having a variable number of threads.

First, we allocate once at startup time a memory block on the GPU with the usual CUDA allocation method. This block has the size of all rx/tx descriptor rings and of the *mbuf* for storing packets. Note that the *mbuf* must have enough packet slots for all descriptor entries in all rx rings plus packets currently processed by the network function and packets under transmission. If we assume 4 rx descriptor rings, each

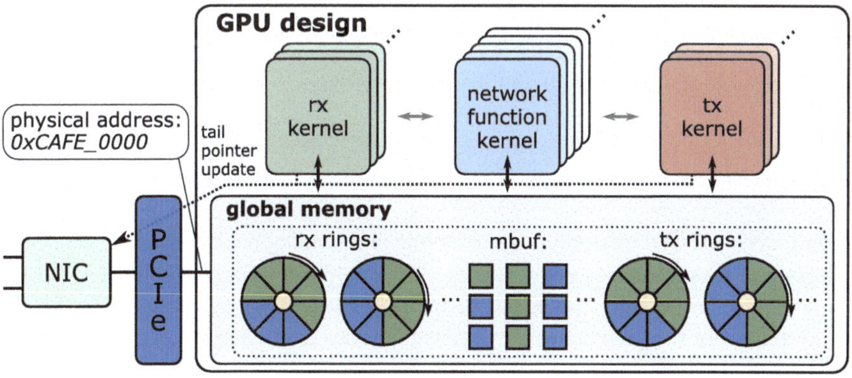

Fig. 3.5 Software structure within GPUs to enable *host bypassing*. The global memory can be accessed by the GPU software kernels but also directly from the NIC

with 256 slots, a total of $> 1,024$ slots is required. However, GPUs provide sufficient global memory, wherefore we allocate in our experiments 16 times as much memory as the descriptor rings have slots.

The newly allocated memory can be accessed from within a CUDA kernel running on the GPU and also from the control process running on the CPU by virtual memory pointers and DMA helper functions. Indeed, physical addressing of this memory by the NIC is not yet possible. For this, the memory must be exposed to a physical address and pinned to it, making it directly accessible from the NIC. The pinning is of tremendous importance as the physical address should be static as a movement would not be noticed by the NIC. For this, we must develop, compile, and load a small kernel module, allowing us to pin and expose the allocated memory. This process is comparable to the NVIDIA GPUDirect communication, and we followed the guidelines of the GPU vendor while developing this module [9].

The kernel module provides an API function *pin_memory*, consuming the virtual address and length of the allocated memory and the GPU's PCIe location (bus and device ID). The CUDA host thread, initializing *host bypassing*, calls this function after the allocation and receives the physical address of the pinned memory as a return value. With this value, the physical addresses of the rx and tx descriptor rings are computed and written into the NIC. Further, this offset is required to compute the physical addresses of memory slots in the *mbuf* memory, which are written in the rx and tx descriptor ring entries. The complete *host bypassing* initialization process is described in Sect. 3.1.3. Let's assume the physical address region is located at the address $0xCAFE\_0000$, the pinned memory region might start at $0xCAFE\_8000$ as the GPU internal offset is $0x8000$. This internal offset and the physical address of the GPU in the system are dynamic and must be determined at run-time.

Modern GPUs provide a complex hierarchy of memories and memory caches in addition to the global memory. Therefore it must be ensured that the *host bypassing* approach allocates the shared memory in the global memory, and memory caches are disabled for all read and write operations on this region. Otherwise, memory inconsistencies could occur between the software kernels running on the GPU and the NIC reading and writing the same memory. For this, we declared all operations on the rx/tx descriptor rings and the *mbuf* as `volatile` operations, enforcing the operation to be performed on the global memory only.

Parallel Packet Handling

We initialize two CUDA streams for receiving and sending packets, each executing a dedicated CUDA kernel, as shown in Fig. 3.5. In addition, a *network function kernel* can be started to perform the actual network function to be accelerated. However, this kernel is not the focus of this work, as the focus is on the packet I/O with good performance.

One or multiple descriptor rings can be used for receiving and sending packets. In the case of an incoming packet at the Ethernet port of the NIC, a hash value is computed based on a couple of header fields. Next, the NIC determines one of the *rx rings* based on this hash value. The remaining receive process for each ring follows the usual procedure of poll-mode drivers as introduced in Listing 3.1. In order to support multiple receive threads in parallel, the *rx kernel* has a 1-dimensional block ID, indicating to which ring this kernel belongs. The descriptor rings are located within the global memory in ascending order. Therefore, the kernel can easily compute the memory address, i.e., a memory pointer, of its descriptor ring at run-time by the block ID, known memory base address, and the descriptor ring size.

Analogous to the receiving process, sending packets can be done by multiple CUDA threads in parallel within the *tx kernel*. For this, any packet distribution over the threads is viable, and no classification must be performed, as long as no single thread is overloaded. For our experiments, we realized a one-to-one mapping of ingress and egress threads, e.g., a packet received by the rx thread 3 would be sent by the tx thread 3. As long as no huge amount of packets is generated within the GPU, e.g., for multicast applications, the load distribution at the tx side is sufficient and inherited from the NIC rx classification.

For the execution of a network function, we propose a similar concept: a *network function kernel* is started consisting of multiple threads. This number should be at least the same as the number of *rx/tx kernel* threads. In the case of more threads, the *rx kernel* performs a load distribution of incoming packets on all *network function* threads being assigned to it. For example, 64 *network function* threads may be managed by 8 *receive* threads. Note that no network function on the GPU is developed within this work, as we focus on packet I/O only.

The communication between the three kernels running in parallel, realized as CUDA streams, poses challenges. It is required to realize an inter-stream communication via the global memory, suffering under high latency. An alternative represents the communication between threads within a single kernel, accessing a so-called "shared memory" with significantly better performance. Therefore, the investigation of different parallelism mechanisms within the GPU would be a promising continuation of this work, e.g., realizing packet reception, sending, and processing within the same CUDA kernel but with higher parallelism.

Memory Management

The memory region in the global memory, including the rx/tx descriptor rings and *mbuf*, is allocated once at startup by the initialization process. However, within the *mbuf* an allocation procedure is required. As a shared memory buffer would require synchronization between the rx threads, we realized a *mbuf* slice for each receive thread. Assuming an uneven distribution of the packets over the rings, this requires a little bit higher memory utilization. However, GPUs provide sufficient global memory resources, realized in *DDR5* memory. Our allocation mechanism allows each *mbuf* packet slot to be in one of the following three states: (1) free, (2) allocated by the *rx kernel* for future incoming packets, or (3) under processing by the network function or *tx kernel*. The *rx/tx kernels* can allocate and free memory. Consequently, this zero-copy implementation allows a packet to be stored at the same memory address in the global memory from the receiving, during the network function processing, until sending.

3.1.6 Related Work on PCIe Hardware Accelerator Integration

Provisioning and interconnection of hardware accelerators in PCIe environments have been discussed in related work before.

Nobach et al. presented a framework for elastic provisioning for hardware-accelerated virtual network functions [35]. In contrast to traditional Network Functions Virtualization (NFV), they suggest to shipping network functions as a software bundle with different implementations, including a legacy CPU implementation and various implementations for offloading performance-critical parts of the network function on hardware accelerators, namely FPGAs, GPUs, and Network Processing Units (NPUs). When deployed on a server, the available hardware accelerators are checked, and the network functionality may be offloaded to an accelerator. A similar approach was presented almost at the same time, both in 2015, by *Bronstein et al.* [6]. However, both approaches utilize only state-of-the-art packet I/O capabilities and would enormously benefit from the concepts presented in this section.

Utilizing the PCIe peer-to-peer capabilities in general for accelerating program execution was discussed multiple times in related work. *Thoma et al.* presented a framework, named $FPGA^2$, for direct communication between FPGAs and GPUs [42]. Similar to our approach, they benefit from fewer data transfers in the system and, by that, increased throughput at lower latency. A similar approach of the direct interconnection of FPGAs and GPUs was presented by *Bittner et al.* in 2014 [4]. On the example of a computer vision application, they demonstrated that the approach could lower the latency significantly. The authors mention that a peer-to-peer data transfer requires the willingness of GPU vendors to enable these features, as they "are presently hidden behind black-box driver code." Almost one decade later, only a few premium GPUs support this feature, and cheap consumer cards still cannot be used. However, we are convinced that this is caused by commercial reasons only and not a technical limitation.

Markussen et al. investigated the capabilities of sharing PCIe devices between multiple servers. For that, they utilize a PCIe network, interconnecting the servers, and physical address translation capabilities of the servers, i.e., by the I/O Memory Management Unit (IOMMU). The presented approach of the authors also leverages PCIe peer-to-peer capabilities and experienced performance benefits from avoiding unnecessary copies to the systems main memory [29, 31]. In a follow up work, they extended this approach to operate on non-volatile memories leveraging NVIDIA GPUDirect capabilities, i.e., accessing Solid State Drives (SSDs) directly from the GPU [9, 30].

3.2 Internet Service Creation on Programmable Hardware

According to the first research challenge of this thesis, network functions must provide high flexibility and functionality while providing high performance. In this section, we will focus on this in the context of Internet service creation. More in detail, the focus is on the point of subscriber termination, i.e., the *access edge* of residential and mobile Internet access networks. In particular, as introduced in Sect. 2.2, existing approaches suffer either under limited flexibility or relatively low performance.

In the subsequent Sect. 3.2.1, we discuss and analyze the requirements for Internet service creation. In Sects. 3.2.2 and 3.2.3, we present our solution approaches for residential and mobile Internet access, leveraging programmable hardware. Last, Sect. 3.2.4 introduces a concept for flexible packet queueing, shaping, and QoS-enforcement, building upon FPGAs.

3.2.1 Functional Requirements Analysis

The underlying protocols and mechanisms for residential and mobile Internet service creation are different; however, they build upon similar concepts. In Fig. 3.6, the similarities between both approaches are presented visually.

First, we discuss the *subscriber tunneling* from the customer premises equipment, i.e., the Residential Gateway (RG) for residential and the User Equipment (UE) for mobile subscribers, to the access edge. For this, it is required to provide an isolated tunnel for each subscriber, implying a large number of parallel tunnels to be terminated at the access edge. For residential access networks, typically, the Point-to-Point Protocol over Ethernet (PPPoE) protocol is used to enable per subscriber data encapsulation. The tunnel is terminated by the Broadband Network Gateway (BNG). In 5G networks, this tunnel is realized between the User Plane Function

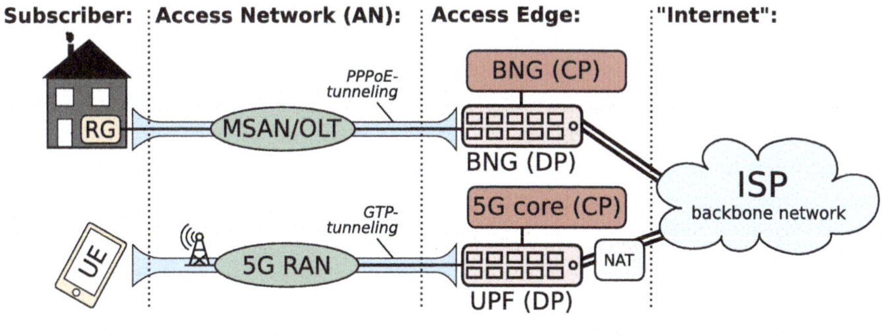

Fig. 3.6 Subscriber tunneling and termination in Internet service creation for residential and mobile subscribers

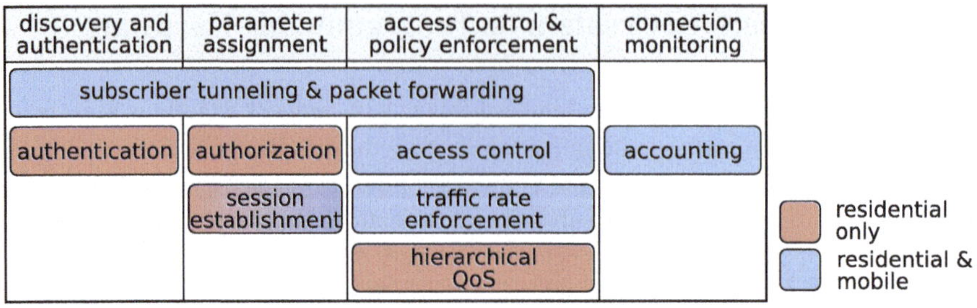

Fig. 3.7 Summary of functional requirements in Internet service termination for residential and mobile subscribers at the BNG/UPF. Figure derived from: [23]

(UPF) and Radio Access Network (RAN) by the GPRS Tunneling Protocol (GTP) in mobile access networks. For the radio interface, other 5G-specific protocols are used. However, the end-to-end tunnel characteristics between UE and UPF are not affected.

Second, the Access Node (AN) is responsible for forwarding packets to and from the subscriber. In both scenarios, the AN can enrich upstream packets with additional metadata, i.e., the AN line ID for residential access or the RAN cell to which the UE is attached. This metadata information can be used during authentication and authorization by the BNG or 5G core control plane (CP), respectively.

Besides subscriber tunneling and packet forwarding, further functional requirements exist in Internet access networks, as shown in Fig. 3.7. It is noteworthy that the UPF is only responsible for data plane traffic termination in 5G access networks. The authentication, authorization, and session establishment request messages are sent directly from the RAN to the Access and Mobility Management Function (AMF) of the 5G core. The UPF is involved in the session establishment by applying the flow rules from the 5G core, more precisely from the Session Management Function (SMF).

In contrast to the UPF in the highly disaggregated 5G architecture, the BNG functionality includes the control functionality for authentication, authorization, and session establishment. Further, the BNG must provide *hierarchical QoS*, including traffic shaping, in the downstream direction for each subscriber. For the UPF, *traffic rate enforcement* for each established subscriber session is sufficient, and no hierarchical dependencies must be considered. In both scenarios, traffic rate enforcement in the upstream direction is needed.

In addition to this, the access edge must perform *access control* checks on each packet. These checks include the validation of the sender IP address, named *anti-spoofing* filtering. Otherwise, a malicious subscriber could send network packets into the Internet with a sender IP address not belonging to him, e.g., to perform a cyber attack. Similarly, it must be avoided that a subscriber can send network packets in a higher QoS class than they belong to, e.g., misusing the Voice over IP (VoIP) traffic class.

Last, accounting for the traffic of each subscriber is essential for two reasons: First, this measurement data is essential for network monitoring, failure detection, and network planning purposes. Second, volume contracts require to slow down a subscriber dependent on their accumulated bandwidth usage or to charge additional fees.

To summarize, the requirements on residential and mobile Internet access edges have several overlaps. Therefore, the two approaches presented in the following sections have high similarities.

3.2.2 P4-Based Broadband Network Gateway

In this section, we present how a Broadband Network Gateway (BNG), the access edge for residential Internet subscribers, can be realized flexibly but still with high performance, utilizing programmable hardware. This section excludes the traffic shaping in the downstream direction and focuses on the header processing. The aspect of hierarchical QoS and traffic shaping, in general, is discussed later in Sect. 3.2.4 for residential and mobile Internet access creation. The presented approach presented in this section is available open-source on GitHub.[1]

Packet Header Decoding

In the following, we assume the header stack as shown in Fig. 3.8. Upstream packets, i.e., data from the subscriber to the Internet backbone, are encapsulated by the RG in a PPPoE tunnel. In addition, the RG adds a Virtual LAN (VLAN) tag to identify the type of service, and the Ethernet source and destination addresses are updated. In the next step, this packet is forwarded to the access node, which adds a second VLAN header to the packet. This second VLAN header represents the line ID within the access node and can be used for authentication. Finally, the BNG receives this packet, including all added headers, and must interpret them.

Similarly, the packets arriving from the backbone site consist of a custom header format, specifically two stacked Multiprotocol Label Switching (MPLS) headers. MPLS is a common protocol in Internet Service Provider (ISP) backbone networks, relying on source routing mechanisms. In both directions, the BNG must parse the incoming packets correctly to allow further processing.

However, understanding these protocols, which are not even supported by commonplace SDN hardware [34], is challenging. Therefore, the programming language P4 and corresponding hardware architectures provide a novel degree of freedom wherefrom we benefit in this work.

[1] https://github.com/opencord/p4se.

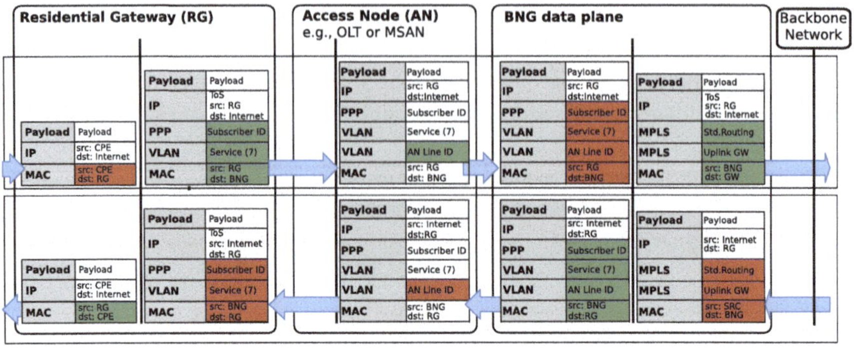

Fig. 3.8 Packet header-centric system design of a residential access network. Red headers will be removed, green headers will be added. Figure derived from: [23]

31	27	23	15	0
version	type	code	session ID	
length			protocol	

Fig. 3.9 Point-to-Point Protocol over Ethernet (PPPoE) header format for data packet encapsulation (PPPoE Session Stage), according to: [7]

The main challenge of residential Internet access termination with commodity SDN-hardware is the PPPoE protocol [34]. A PPPoE packet is transported in an Ethernet frame, and the header can be identified by a known *EtherType*, $0x8863$ and $0x8864$, for PPPoE control and data packets. The structure of PPPoE data packets, formally named "PPPoE Session Stage," is shown in Fig. 3.9. Analogous to the P4 example in Appendix A.1, we define this header format (and all additionally required headers) in P4. The parser state machine, also part of the P4 program, contains a specific rule to extract this PPPoE header in the case of the dedicated EtherType.

At the access side of the BNG, the following packet header to be parsed can be determined by the appropriate next header fields of the currently parsed header, e.g., the *EtherType* in the Ethernet and VLAN headers or the PPPoE *protocol* field. In contrast, packets arriving at the backbone side consist of two MPLS headers, which do not indicate the encapsulated packet type. To determine the next header in this case, we introduce a lookup table in the parser, building upon the P4 construct of *parser value sets*. By this, the next protocol type is determined by the MPLS label, for example, all flows with the label 42 comprise of IPv4 packets. As the network operator is aware of its own MPLS routes, the next packet header, i.e., *IPv4* or *IPv6*, can be parsed.

After the parser completes extracting all required headers, they pass through the switch pipeline as a P4 packet header vector. Multiple tables and actions can be applied to realize the desired BNG functionality based on this vector. The deparser assembles the packet headers from the vector to a packet at the end of the pipeline.

For this, all valid headers are deparsed in a given order. Note that the packet header vector, including the set and kind of headers, can change while being processed in the pipeline.

P4-Based Pipeline Design

In the following, we discuss the processing of the parsed headers in detail. Note that many packet header vectors may be processed parallel by the switch pipeline in different stages. Therefore, it is mandatory that every pipeline stage has a guaranteed limited processing time. Thus, the P4 program must be written in a way that the compiler can fulfill these timing constraints and available resources of the programmable pipeline, as introduced in Sect. 2.3. The following explanations consider only the case of IPv4 packets for simplicity reasons; however, the processing of IPv6 packets functions analogously.

Figure 3.10 depicts the control flow of the P4 program, realizing the BNG functionality. Later, a compiler can map this program on the P4-programmable match-action pipeline resources of the underlying hardware. In general, the proposed pipeline design can be divided into three parts: (1) a general pipeline for packet classification, (2) a downstream pipeline, and (3) an upstream pipeline.

The **general** classification part of the pipeline is colored in light-yellow in the figure and consists of two tables with the indices 0 and 1. First, the table *t_cptap_outer_ethernet* determines, based on the Ethernet destination address and etherType, i.e., the type $0x8863$, whether it is a packet for the BNG control plane. In that case, the packet will be forwarded to the control plane, and no further processing in the P4 hardware is done. This rule applies to all packets during authentication and authorization but also for periodic link control echo requests and responses, used for monitoring the connection state. Similarly, packets can be injected from the control plane, which is not shown in the diagram. In all other cases, the pipeline proceeds with the table *t_usds*, responsible for determining if the ingressing packet belongs

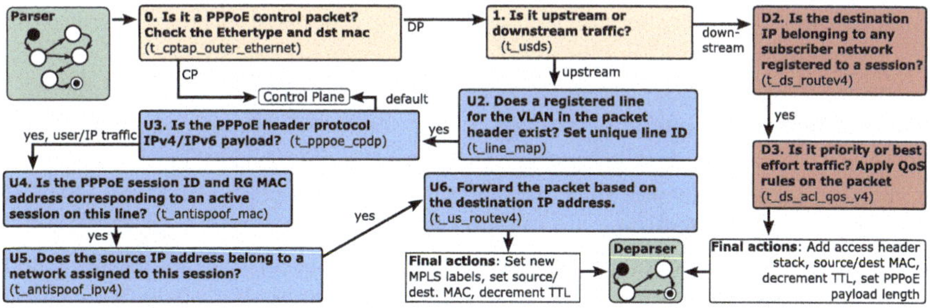

Fig. 3.10 Control flow of the P4 pipeline for realizing BNG functionality. The pipeline can be separated into a general part (yellow and green), an upstream pipeline (blue), and a downstream pipeline (red). Figure derived from: [23]

to an access or backbone switch port. This information is used to select if the packet should be processed by the upstream or downstream pipeline. From a functional side, this could also be done by checking the valid packet headers. However, checking the ingress port is also a security mechanism. Otherwise, a malicious subscriber could send a valid double-tagged MPLS packet to the BNG, interpreted as a packet from the trustworthy backbone side.

In the **downstream** direction, packets arrive at a port attached to the MPLS-routed backbone network of the ISP. Therefore, incoming packets at this port can be assumed to be more trustable, assuming security mechanisms on all ingressing ports of the backbone network. The first applied table for downstream traffic is *t_ds_routev4*. Based on the destination IPv4 address (or IPv6 address range), we determine the corresponding subscriber, including its subscriber ID and line ID. This subscriber information is used in the next step to encapsulate the packet in a PPPoE tunnel and two stacked VLAN headers, as discussed before. In addition to the PPPoE session ID the length of the packet is required for encapsulation and can be determined from the length field of the ingressing IPv4 packet. As modern switches start packet processing as soon as all required header fields are received and parts, the tail of the packet may still be under transmission, and therefore the length of the received Ethernet frame cannot be referenced. If the subscriber's IP address is unknown, the packet is dropped and may be reported optionally to the control plane for monitoring reasons.

Second, the table *t_ds_acl_qos_v4* is applied. Mainly based on the sender IP address, the QoS class of the packet is determined. As all non best-effort traffic classes, e.g., VoIP or live television streaming, have their source within the ISP backbone network, the addresses are known and can be used. QoS class identifiers set by either subscribers or third-party data center operators are not trustworthy and should be overwritten. Otherwise, this invites abuse.

In the **upstream** direction, first, the table *t_line_map* determines the unique *line ID* of the subscriber based on the access node line ID, provided as a VLAN tag, and the belonging access node. Note that this involves no information provided by the subscriber, e.g., the PPPoE session ID, which could be manipulated. In the next step, table *t_pppoe_cpdp* determines based on the PPPoE protocol field if the payload is a valid IPv4 (or IPv6) header and the packet can be processed further. Otherwise, in the case of an unknown PPPoE protocol type, the packet is sent to the control plane.

The next two tables, *t_antispoof_mac* and *t_antispoof_ipv4*, check if the previously computed unique line ID maps to the sender's Ethernet and IPv4 source addresses. This prevents a subscriber from impersonating another subscriber or anonymous packet injection into the Internet, named *spoofing* [14]. As an RG of a subscriber typically has only one Ethernet address but probably multiple IPv4 and IPv6 addresses, this functionality is split up over three tables (for Ethernet, IPv4, and IPv6 source addresses) to lower the total resource utilization. If a subscriber sends a packet with an invalid source address, the packet is dropped in the data plane.

Last, Table *t_us_routev4* assigns a route to the packet based on its destination address (IPv4 or IPv6). For this, two MPLS labels and a new Ethernet destination address are added to the packet, analogous to a conventional MPLS edge router that forwards packets through the MPLS network. All the access headers, including the two VLAN headers and the PPPoE header, are removed. After this processing, upstream packets are deparsed and sent out to the backbone side.

Subscriber and State Management

The presented pipeline design is optimized for three target platforms: (1) the P4-NetFPGA, (2) P4-programmable SmartNICs manufactured by Netronome, and (3) Intel Tofino-based network switches. Even though the data plane design is similar for all three platforms, the control plane interfaces were not unified at the time of this work. Later, in Sect. 5.2.1, we provide evaluation results for these three platforms, highlighting the pros and cons of each. To provide a unified state configuration API during evaluation, allowing to set up new subscriber session and a general platform configuration, we build a simple python library. This library allows adding/removing subscribers and configuring access and backbone ports. However, note that the hardware abstraction is not the focus of this work and only a means to an aim.

3.2.3 P4-Based User Plane Functions in 5G Networks

The User Plane Function (UPF) in mobile networks is the equivalent of the BNG data plane, which is discussed in the previous section. Note that the UPF is only responsible for the subscriber termination on the data plane, as the radio access network sends all control messages directly to the 5G core control plane. Therefore, the UPF contains less control logic and we can present a fully-fledged prototype in this work, which can be integrated in the remaining 5G core via the unified N4 interface. Note that the 5G core is formed by the control plane network functions and the UPF as the data plane function. Even though the integration in an end-to-end network is an extensive research field, it is not the focus of this thesis and is explained briefly in Appendix A.2.

The differences between multiple programmable hardware platforms are already investigated by the BNG use case and it is very likely that the general characteristics continue to possess their validity for the UPF use case. Thus, in the following investigations, we consider only a single programmable architecture, specifically the P4-programmable Intel Tofino.

In the remainder of this section, we will discuss how concepts and mechanisms from the BNG scenario can be adapted and unified for the mobile 5G access use case, fulfilling the demanding QoS requirements on current and future 5G networks.

General System Design

The UPF system design consists of multiple components and is shown in Fig. 3.11. The system is connected to its environment through three interfaces, *N3*, *N4*, and *N6*, following the 3rd Generation Partnership Project (3GPP) specification [2]. In this work, we present a disaggregated concept within the UPF, serving these three interfaces. The *control plane* component of the UPF implements the *N4*-interface protocol, allowing to add, modify, and remove subscriber sessions.

From the control plane, all deployable flow rules are handed over to the *hardware abstraction layer*, which provides a simplified API to configure ports and manage subscriber sessions. A flow rule may not be deployable if only a subset of the required information is handed over to the UPF from the SMF during subscriber registration. In that case, the state is stored only in the UPF control plane until all required information is present. The simplified API can be used to migrate the presented approach to other programmable platforms, e.g., the P4-NetFPGA or Netronome SmartNICs, as evaluated in this work for the BNG scenario.

Next, the hardware abstraction layer installs flow rules on the hardware. As the current P4-programmable hardware does not offer sufficient resources for per-subscriber traffic shaping, we added an optional QoS co-processor, realized on an FPGA directly attached to the P4 switch. The abstraction layer divides a single incoming request over these two hardware platforms. For example, a new downstream subscriber session would be installed (1) in the P4-programmable ASIC to encapsulate and forward its packets, and (2) in the QoS co-processor, shaping the packets to the maximum allowed bandwidth of the subscriber. The queue assignment is done within the P4 switch and this information is given to the QoS co-processor together with the packet. Details on the QoS co-processor concept are discussed later in Sect. 3.2.4.

Furthermore, the system consists of a software based packet processor, labeled as *slow path*. This software switch is used for replying to address resolution requests, Ethernet address learning, and Ethernet address resolving. All packets which cannot

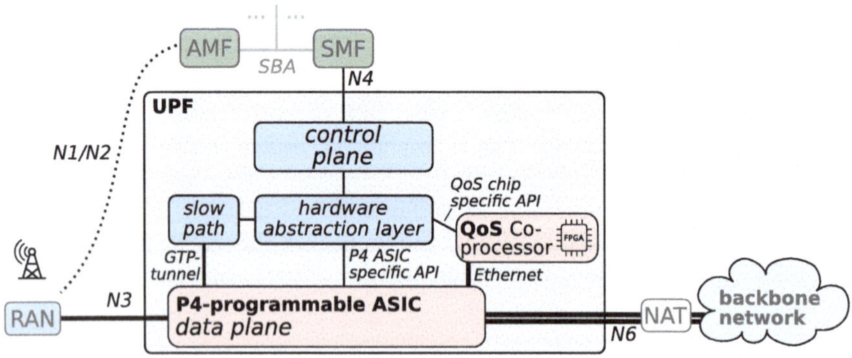

Fig. 3.11 System architecture of the hardware accelerated UPF presented in this work. Data plane components are highlighted in light red, control plane functionality in blue

be processed by the P4-based data plane are encapsulated in a GTP tunnel and sent to this slow path switch. The GTP encapsulation has two advances: First, the slow path part of the UPF can be located anywhere in a data center, reachable by the UPF data plane. This is also the case for the hardware abstraction layer and control plane. Thus, the UPF software stack can be either located in the same box as the programmable hardware or distributed and probably virtualized. Second, the GTP Tunnel Endpoint ID (TEID) is used to inform the slow path processor on which port the packet ingressed the switch.

Besides receiving packets, the slow path switch can inject packets into the data plane on any port of the programmable ASIC. For this, the GTP tunnel ID, set by the software switch, indicates the port on which the packet should be sent by the hardware switch.

The normal case of operation is as follows: First, a subscriber flow for upstream and downstream traffic is installed via the N4 interface. Following this, the subscriber can send IP packets addressed to the backbone network via the RAN on the N3 interface into the UPF. The UPF terminates the subscriber tunnel and forwards the packets to the N6 interface. Analogously, downstream packets arriving at the N6 interface are sent towards the RAN on the N3 interface.

We realized all software components, including the control plane, hardware abstraction layer, and slow path, using the *Golang* programming language. This language provides ideal support for the PFCP protocol due to existing libraries, used at the N4 interface [27]. Further, the integration of the gRPC and GTP protocols can be easily done and the program execution performance is sufficiently high.

Subscriber Data Encapsulation

For each subscriber, more precisely for each "Protocol Data Unit (PDU)" session, at least one tunnel is established between the RAN and the UPF. In case of additional services, e.g., VoIP, multiple tunnels are established in parallel for a single subscriber. The header stack for packet encapsulation, i.e., for tunneling, is shown in Fig. 3.12. Note that the RAN is not the focus of this work, and therefore the protocols of the air interface are not shown.

The packets arrive at the access side of the UPF encapsulated by the GPRS Tunneling Protocol (GTP), a tunneling protocol on top of the UDP/IPv4 or UDP/IPv6 network protocols. The Ethernet, IP, and UDP headers can be parsed in a straightforward manner based on the respective *next protocol* field. However, the GTP protocol can only be assumed based on the provided UDP destination port, typically port 2152 for GTP. Within the GTP encapsulation header, the IPv4/IPv6 packet sent by the subscriber is encapsulated. On the backbone side of the UPF, the packets arrive and depart as conventional packets. In mobile networks, a Network Address Translation (NAT) is typically instantiated between the backbone network and the subscriber for several reasons (compare Sect. 2.2.4 for details). Therefore, we can assume a direct connection at the backbone side of the UPF to the NAT, and no complex routing mechanism is mandatory.

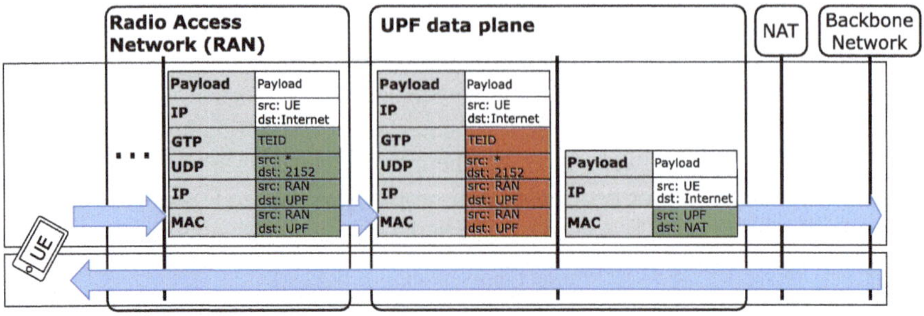

Fig. 3.12 Packet header-centric view on the UPF functionality. Downstream packets are only indicated by the arrow; the added/removed headers are reverse to the upstream direction. Red headers are removed, green headers are added

Most used network protocols on both sides of the UPF are very common, and the parsing process is straightforward. Only the GTP protocol poses a minor challenge, as it is encapsulated in a UDP frame that does not indicate the next protocol type. However, as stated before, the parser state machine can determine the next parsing state based on the UDP destination port. UDP ports are used to differentiate multiple applications running on a single system, e.g., port 2152 for GTP tunneling. Transferring this concept of application differentiation to the hardware, we can state that the parser filters incoming packets based on the destination port for the UPF application running within the switch. Alternatively, to support a different UDP-port or multiple ports in parallel, this value can be set dynamically at run-time by a P4 parser value set. This concept is very similar to the parsing of MPLS packets at the BNG, as discussed in the previous section.

Figure 3.13 depicts the header format of the GTP protocol. The header has a base length of 8 bytes, including the payload *length* and the *Tunnel Endpoint ID (TEID)*. In addition, further optional extensions can occur, indicated by the status bits in the *flags* section of the base header. In our experiments, we observed different GTP header formatting in terms of extensions, and therefore the UPF must be able to interpret all of them. Specifically, the QoS identifier, named *5QI*, is an extension in 5G standalone and thus is not present in older standards. As a consequence, it is currently not implemented in all existing UPF implementations. After extracting the header fields of the base header, the parser checks if at least one of the flags bit is set and proceeds in parsing the optional header extensions. If the *next extension* field of the currently parsed extension is zero (0), the parser stops and hands over the packet to the ingress pipeline. In the downstream direction, the packet deparsing works analogously but is under the control of the UPF, which can decide the exact header format. However, the downstream header format must be specified at compile time of the P4 program, i.e., if an optional extension for the QoS identifier should be added to every packet.

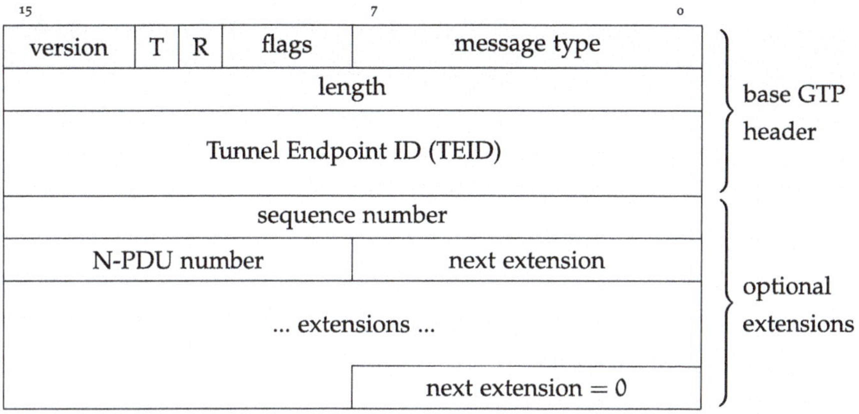

Fig. 3.13 GPRS Tunneling Protocol (GTP) header format according to: [1]. Note that the header visualization has a width 16 bit for readability reasons and the presentation may differ from other literature

Data Plane Pipeline Design

The packet processing part of the UPF consists of a P4 programmable switch and an optional QoS co-processor. In this section, we discuss the internal structure and concepts of the programmable switch, realizing the header processing within the UPF. In the first step, the incoming packets are parsed, as discussed in the previous section. Next, the packets walk through the programmable match-action pipeline of the switch, consisting of multiple lookup tables and actions, as shown in Fig. 3.14.

The pipeline, described in the P4 programming language, can be separated into four parts, a *general* packet processing, an *upstream pipeline*, a *downstream pipeline*, and a *slow path* injection pipeline.

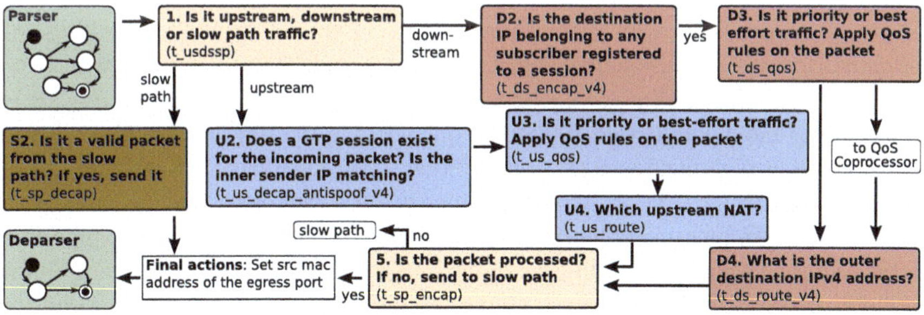

Fig. 3.14 Control flow of the P4 pipeline realizing the UPF data plane functionality. The pipeline consists of a general part (yellow and green), an upstream pipeline (blue), a downstream pipeline (red), and a slow path injection pipeline (beige). Figure derived from: [23]

In a first step, ingressing packets are processed by the table *t_usdssp* of the **general** pipeline. This table is used to determine by which pipeline the packet should be processed next. For this pipeline assignment, we use the ingress port of the switch; however, in the case of programmable devices with less ports other packet characteristics, e.g., the destination IP or available packet headers, can be used. Next, the packet is handed over to one of the three pipelines.

The **upstream** pipeline consists of three tables. First, the table *t_us_decap_antispoof_v4* and the belonging action decapsulate the incoming GTP packet. For this, the table uses the GTP tunnel ID and the IPv4 source address of the inner, encapsulated IP packet as input. As each subscriber session has an assigned IP address, this table combines the tunnel termination and anti-spoofing detection. If a subscriber sends a packet with a spoofed sender IP address, this table will not match, and the packet is discarded. If a flow rule matches, i.e., the incoming packet belongs to an authenticated subscriber, the action of the table invalidates all encapsulation headers, as shown in Fig. 3.13.

Next, the optional table *t_us_qos* is applied. This table matches the same flow ID and applies a metering action on the packets according to their configured maximum upstream bandwidth. In contrast to traffic shaping in the downstream direction, the metering requires no packet queueing, and packets are either forwarded or dropped.

The last upstream specific table is *t_us_route*. This table and the belonging action determine the egress port and next-hop Ethernet destination address, i.e., to which Network Address Translation (NAT) the packet should be sent. This table can load-balance the upstream packets between multiple NAT instances that are common in mobile access networks. But even if only a single NAT is used, this table strongly reduces the action data width of the preceding table, as only a single table entry is required to forward the packets of all subscribers attached to this UPF. Finally, the packet will be processed by the general part of the pipeline and sent out.

The **downstream** pipeline has an inverted behavior compared to the upstream pipeline. First, the table *t_ds_encap_v4* is responsible for the GTP encapsulation of incoming packets. For this, the IPv4 destination address of the incoming packet is used as match input. Suppose a registered subscriber belongs to this IP. In that case, the packet is encapsulated in the corresponding GTP tunnel, using the IP address of the connected radio access network and the GTP tunnel ID of the subscriber.

Next, if a QoS co-processor is connected to the P4 switch, the table *t_ds_qos* is applied. This table determines the unique queue ID of the subscriber and sends the packet together with the queue ID to the FPGA. The processing within the FPGA is discussed later in Sect. 3.2.4. As soon as the packet is sent back from the FPGA to the P4 switch, the pipeline proceeds with the third table of the downstream pipeline. To allow this, we added a fourth case in the initial classification table of the general pipeline part. Figure 3.14 shows a simplified version of the FPGA integration to improve the readability. In the actual design, the packet enters and leaves the P4 switch twice, including parsing and deparsing. If no QoS rule is installed for this subscriber, the packet is handed over directly to the next table, *t_ds_route_v4*.

Last, Table *t_ds_route_v4* determines the egress port and Ethernet destination address of the packet based on the outer IPv4 destination address, i.e., the addressed 5G base station.

The third pipeline of the presented design is the **slow path** packet injection. Packets to be sent by the software switch arrive at the P4 switch in GTP encapsulated packets. These packets are decapsulated and sent to the egress port specified in the GTP tunnel by the software switch. Note that the encapsulated packets also include an Ethernet header, and therefore no destination address must be determined in the P4 switch. This is intended by design, as the slow path is used to request unknown Ethernet addresses and should be able to send out any packet on any port.

At the end of the upstream and downstream pipelines, all packets are processed again by the **general** part of the pipeline. The table *t_sp_encap* verifies if a packet was successfully processed by one of the pipelines. If not, the tables action encapsulates the packet and marks it to be sent to the slow path switch port. Finally, the P4 pipeline sets the Ethernet source address of the packet corresponding to the egress port, and the packet is handed over to the deparser for sending.

End-to-End UPF Integration

The presented UPF design provides all required functionality to be integrated into a 5G standalone network, including a standard-compliant 5G core and Radio Access Network (RAN). In Appendix A.2, we describe the end-to-end testbed used to verify the functional correctness of this prototype. In total, we tested three different RANs from different vendors and the free5gc open-source 5G core [11]. We observed multiple issues in the open-source 5G core during integration and testing, which we repaired and contributed to the open-source project. The proposed design (with and without QoS Co-processor) works as expected to the best of our knowledge. However, the presented prototype's main purpose is to demonstrate hardware acceleration in mobile access networks, and therefore we do not aim to build a commercial product.

3.2.4 FPGA-Based Traffic Shaping

In the two sections before, Sects. 3.2.2 and 3.2.3, we discussed how residential and mobile Internet access could be realized with programmable hardware. While the functional header processing can be realized with existing programmable network switches, they do not provide sufficient resources for per-subscriber traffic queueing. If we assume 35, 000 residential subscribers, each having 4 QoS classes, a total of 140, 000 parallel queues are required. Indeed, most current network switches, including the P4-programmable switch utilized in this work, offer only a few queues, e.g., 8 queues per egress port. Therefore, subscriber isolation and per-subscriber traffic shaping, a requirement stated in Sect. 2.2, is not possible with current P4 switches. However, we introduced the concept of a QoS co-processor for both scenarios, res-

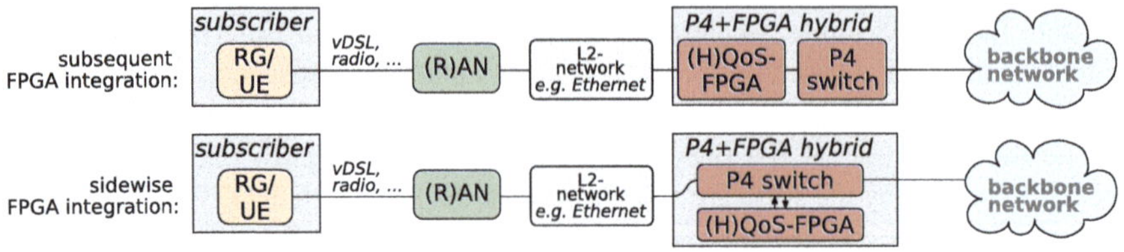

Fig. 3.15 Integration concepts of an FPGA-based QoS co-processor with P4-programmable switches. Figure derived from: [23]

idential and mobile Internet service creation. In the following, we will discuss this co-processor in detail.

We decided to realize the QoS co-processor within Field Programmable Gate Arrays (FPGAs). FPGAs provide the potential to realize almost every digital circuit on bit-level, including a QoS-aware traffic shaping system. Though, as stated in the background Sect. 2.3, they have a significantly lower clock frequency, wherefore the maximum achievable throughput is lower, i.e., in the range of multiple hundred Gbit/s. Yet, this is sufficient to fulfill the bandwidth demands of current and near-future access edges for residential and mobile Internet subscriptions. In contrast to manufacturing Application Specific Integrated Circuits (ASICs), development time and costs are significantly lower for FPGAs. Note that the presented design could also be realized within an ASIC; however, this would probably not be economical for a product series of multiple thousand devices of this niche functionality compared to commodity data center switches. Further, in contrast to ASICs, FPGAs allow a reconfiguration of the chip after deployment, i.e., an update can be deployed on the access edge devices, providing new functionality or rectifying faulty behavior.

Figure 3.15 depicts two concepts of integrating an FPGA into the data plane. Note that the FPGA-based QoS co-processor is only used to shape downstream traffic.

(1) The FPGA can be attached to the subscriber side of the P4 switch in the data path. By this, the access ports of the access edge, i.e., the BNG or UPF, are realized in the FPGA. On the one hand, this allows realizing many different link-layer technologies, including almost any Ethernet standard from 10 Mbit/s to 400 Gbit/s. But also, specific link-layer technologies in access networks, e.g., XGSPON in Fiber to the Home (FTTH) access networks, can be realized on the same chip. Theoretically, the access node and access edge can be located in the same hardware platform. On the other hand, multiple link-layer technologies must be realized in parallel in the FPGA design to support different generations of access network nodes. This includes multiple Ethernet speeds, but auto-negotiation and link training technology are also mandatory. Furthermore, the FPGA must route the packet, i.e., sending it to the appropriate egress port where the subscriber is attached.

(2) An alternative to this is the sidewise FPGA integration. Here, the packets arrive at the backbone side of the switch. After being processed, they are sent to the FPGA. From there, after being dequeued, the packets are sent back to the P4 switch. In this sidewise integration concept, the FPGA must support only a single link technology,

e.g., 100 Gbit/s Ethernet, and neither auto-negotiation nor link training must be supported, as the FPGA ports are not exposed to the outside of the hardware platform. Though, the packet traverses twice the P4 switch, halving the switch bandwidth and adding additional latency to the packet. As the maximum bandwidth of the P4 switch is roughly ten times higher than for the FPGA and the latency of one switch pass-through is below 1 μs, we assess this disadvantage to be minor. All dequeued packets on the FPGA are sent back to the P4 switch, which then determines the egress port, lowering the complexity within the FPGA compared to the subsequent integration concept.

Based on this assessment, we focus on the sidewise FPGA integration in the following. Nevertheless, the presented concept for traffic shaping can also be applied for subsequent FPGA integration.

Hierarchical QoS in Residential Access Networks

In residential access networks, a Hierarchical Quality of Service (HQoS) must be applied to downstream packets and is discussed in the following. HQoS is required for legal reasons, as (almost) zero packet loss must be ensured between the subscribers and the access edge, i.e., the BNG. By this, the counted and accounted data traffic can be ensured to be delivered to the subscriber. Note that this does not apply to mobile access networks, where a simple traffic shaping for each subscriber is sufficient.

Figure 3.16 depicts the access topology of residential access networks on the right side, as already known from the background Sect. 2.2. The mapping of the network nodes on the right side into the hierarchical limits is shown on the left side.

For example, the Multi-Service Access Node (MSAN) in the presented figure has a total bandwidth of 1 Gbit/s towards the BNG, and 100 subscribers with each 100 Mbit/s Internet contracts are connected to this MSAN. In general, it is very unlikely that all subscribers utilize their access line fully. If all subscribers try to

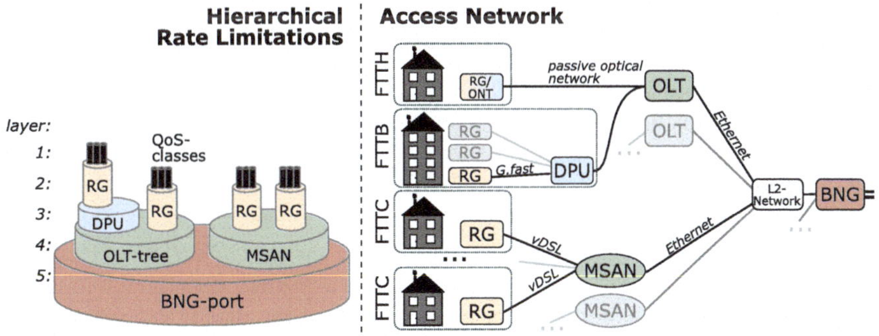

Fig. 3.16 Residential access networks with hierarchical structured network nodes. The topology on the right side is mapped into the layered, hierarchical rate limits on the left side. Figure derived from: [23]

use their total available bandwidth, a bandwidth demand on the MSAN of 10 Gbit/s occurs and cannot be satisfied. This is called *oversubscription*, a typical concept in access networks, viable by traffic statistics. However, in the seldom cases of overloaded network nodes, this would cause packet loss between the BNG and the subscriber, which is not allowed. To avoid this, HQoS must be applied in the BNG. The HQoS scheduler is aware of all bandwidth limitations in the hierarchical access network and can shape the packets accordingly. By that, packet loss still occurs; however, in the BNG before the packets were counted. In addition, the BNG is aware of the different QoS traffic classes, e.g., best effort and VoIP traffic, and can prioritize a high priority packet of subscriber A over a packet of subscriber B with a lower priority.

This mechanism is called HQoS-aware scheduling and must be realized in residential access edges. Note that the QoS-aware prioritization between subscribers is not exactly specified and is only a soft requirement; however, the hierarchy awareness must be fulfilled. In the following of this work, we present a single FPGA design with two exchangeable schedulers, one fulfilling the HQoS demands of residential access networks and a simple scheduler for the mobile use case. The remaining parts of the design are identical.

Memory Technologies for Packet Buffering

The FPGA-based co-processor for traffic shaping realizes a memory-intensive function in the overall system, as every packet to be shaped must be stored by the FPGA. For this, a memory is required, either inside the FPGA or as an external, attached memory. In general, two memory technologies can be considered: Static Random Access Memory (SRAM) and Dynamic Random Access Memory (DRAM).

SRAM memory cells build upon transistors, preserving a stored bit in an electrical circuit. The manufacturing process of transistors can be combined with the digital logic of FPGAs and ASICs. Therefore, SRAM can be realized on the same chip. Nowadays, FPGAs typically provide many internal SRAM-based memory blocks that can be used for the hardware design, which is described in a hardware description language. However, their storage capacity is limited. As SRAM is available in blocks, they are often named *block RAM* in the context of FPGAs.

In contrast to this, **DRAM** technology stores data in tiny capacitors, representing a data bit by their positive/negative charge. This technology can be used to create significantly larger memories. However, DRAM cannot be realized on the same chip as the packet processing logic due to different manufacturing processes. Therefore, DRAM is typically co-located to processing chips, known as DDR3/DDR4 memory. The memory bandwidth of DRAM memory is typically lower compared to internal SRAM, caused by the lower memory clock frequency, the data bus width and the physical characteristics of capacitors. Further, the access latency is significantly higher, as the reading of a single bit requires measuring the electrical load within the corresponding capacitor, which requires some time. In addition, these capacitors lose their charge and must be refreshed frequently, leading to a sporadic non-deterministic behavior of the memory.

According to Perino et al., the access latency of SRAM is 0.45 ns, and for DRAM 55 ns, i.e., around $100x$ higher [39]. Further, they stated in 2011 that the maximum achievable memory capacity of SRAM is $\sim$ 26 MB and for DRAM $\sim$ 10 GB. The cost of one MB SRAM memory is around $1000x$ higher than a realization with DRAM, and the power consumption of SRAM is also higher. Note that these numbers are from 2011; however, the postulated differences between these technologies still have their validity [28]. Only the achievable memory capacity has increased for both technologies but is still in a similar range. Nowadays, with SRAM memory on special-purpose ASICs, one can realize a memory of $<$ 100 MB with high effort. In FPGAs, achievable storage capacities are significantly lower.

A very simple approach to determine an upper bound for the required memory is as follows: Let us assume $35,000\,subscribers$, each having a maximum bandwidth of 100 Mbit/s and a queue depth of 50 ms. A 50 ms queue delay seems to be sufficient for most flows in the Internet, as determined by a 24 hours measurement in the access network of Deutsche Telekom (compare Appendix A.3 for details). Then, we can determine the required total buffer size B by:

$$B = 35,000\ subscriber \cdot 100\ \text{Mbit/s} \cdot \frac{1\ \text{byte}}{8\ \text{bit}} \cdot 50\ \text{ms} = 21.875\ \text{GB}$$

However, this assumption is not realistic as (1) not all subscribers will utilize their access line in parallel, and (2) the backbone bandwidth of the access edge is lower than $35,000 \cdot 100$ Mbit/s. If we assume a backbone interface speed of $2 \cdot 100$ Gbit/s and a shared memory for all queues, the memory demand is as follows:

$$B = 200\,\text{Gbit/s} \cdot \frac{1\,\text{byte}}{8\,\text{bit}} \cdot 50\,\text{ms} = 1.25\,\text{GB}$$

This estimation is still not accurate as each queue level has a sawtooth-like characteristic, i.e., it is unlikely that all queues are 100 % filled at the same time.

A similar problem was already discussed in the related work before and described in the background of this thesis: In Sect. 2.4, we introduced a formula that describes the required buffer size B for n parallel TCP flows in a single queue. This formula cannot be applied to this problem one-to-one, as the subscriber flows are separated in different queues, and the inter-flow interference is not comparable. However, we can use the following estimation as a lower bound. If we assume $2,000\,subscribers$ utilizing their downstream link in parallel, the applied formula looks as follows:

$$B = \frac{RTT \cdot C}{\sqrt{n}} = \frac{50\,\text{ms} \cdot 200\ \text{Gbit/s}}{\sqrt{2,000\,subscriber}} \cdot \frac{1\ \text{byte}}{8\ \text{bit}} = 27.95\,\text{MB}$$

We can summarize that the required packet buffer memory size cannot be precisely determined but is in the range between 27.95 MB and 1.25 GB. Note that measurements in existing devices or simulations with traffic traces are not meaningful, as the dropping behavior of a queueing system influences the sending rate of the transport layer congestion control and, by that, the input rate of packets. Even the lower bound of this memory range cannot be realized with on-chip SRAM memory in actual FPGAs. Therefore, the utilization of DRAM memory, typically external DDR3/4 memory, is unavoidable and a design basis in the following.

Overall FPGA Design

In both access scenarios, residential and mobile Internet access termination, the requirements on the (H)QoS co-processor are similar: Many queues, e.g., 140, 000 for the residential use case with four queues per subscriber, are required, each being able to build up a queue of up to 100 ms. Furthermore, it would be beneficial for the end-to-end service quality to instantiate an Active Queue Management (AQM) algorithm in the co-processor (compare Sect. 2.4). Last, the scheduler implementation should be exchangeable to realize multiple scenarios, e.g., residential and mobile Internet service creation, with a similar design. In addition to the named scenarios, we claim that the presented concept is even more general and can be used for further use cases with only a scheduler modification, e.g., packet scheduling in industrial computer networks with real-time guarantees for certain traffic classes.

Besides the functional design goals, we identified two requirements from the economic perspective: (1) A modular design allows reusing some components and easy extensibility for new functionality. (2) The proposed concept should be FPGA-type independent, and generic. This means parameters of the design can be reconfigured to support many/few queues, schedulers, and different performance levels, depending on the use case. By this, smaller and by that cheaper FPGAs can be used in scenarios with lower demands. Further, the design should be vendor-independent. Specifically, two big FPGA vendors exist globally, and the concepts should be valid for the programmable hardware structures of both of them.

Based on the requirements, we present a generic packet queueing design, shown in Fig. 3.17. This design builds upon vendor-specific IP-cores only for I/O functionality where this is unavoidable, i.e., the Ethernet functionality, the PCIe control plane interface, and the memory controllers. However, the interfaces for the Ethernet IP-cores are identical for FPGAs from the two focused vendors. For the PCIe control plane interface, the *wishbone bridge* converts the vendor-specific signals into the open-source *wishbone* data bus. For a vendor migration, only the bridge implementation must be exchanged.

The last component of the design, which relies on vendor-specific modules, is the memory interface. Here, we provide a memory abstraction layer, allowing the use of a single memory or multiple memories in parallel to benefit from a higher total memory bandwidth. For a low-performance design, only a single external DDR4 memory may be sufficient. If a higher bandwidth is desired, more memories can be added, including external DRAM-based DDR4 memory but also internal SRAM memories. These SRAM memories are very well suited for storing small network packets, e.g.,

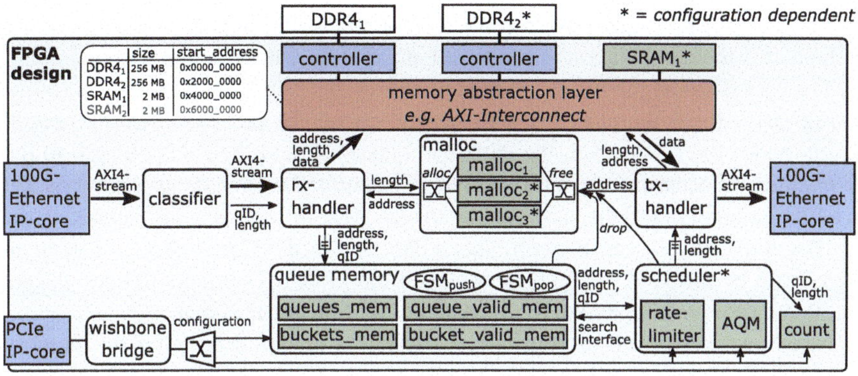

Fig. 3.17 Modular FPGA-internal design of the QoS co-processor. Thick arrows indicate network packets; thin arrows represent metadata and control signals

VoIP data packets, as they probably remain in the FPGA for a short time. Further, DDR4 performs worse for short read and write accesses. As this DDR4 interface is very performance-critical, we optimized the corresponding implementation, i.e., within the *rx-handler* and *tx-handler*, for the AXI data bus, which is used in Xilinx FPGAs. The migration of this interface to Intel FPGAs would either cause some effort or a performance decrease.

In the following subsections, we will discuss some key concepts of the proposed concept in detail.

Packet Data Flow

In the following, we describe the general working principle of the proposed concept. As shown in Fig. 3.18, the P4 switch sends packets to the QoS co-processor with a queue ID, labeled as *qID* in the figure, prepended to the packet. This ID is before the most out header of the packet, i.e., the Ethernet header of the packet, and has a constant length of 4 bytes.

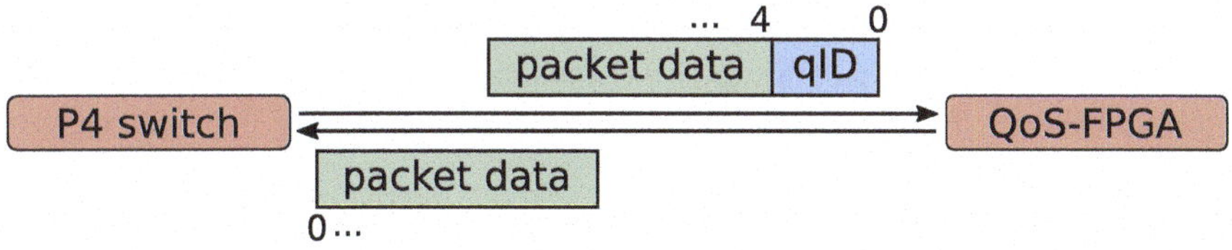

Fig. 3.18 Queue ID metadata transmission from the P4 switch to the QoS-FPGA. The queueing information is prepended to the downstream packet, including all packet headers added by the BNG/UPF

Note that the presented design is not a sequentially executed software algorithm. Instead, it is the combination of multiple hardware modules all running in parallel.

The packet enters the FPGA on the left side at the 100G Ethernet port, as shown in Fig. 3.17. First, the *classifier* cuts off the queue ID, prepended to the packet, and provides the raw packet, the extracted queue ID, and the packet length to the *rx-handler*. To determine the packet length, it is required to store the incoming packet in a FIFO memory until the last byte has arrived. The utilized data bus, defined in the AMBA AXI4-stream specification, has a data width of 512 $bits$, e.g., a 1500 byte packet can be received in 24 clock cycles of the FPGA. The queue ID and packet length are provided in parallel to this data bus as a metadata signal.

Next, the *tx-handler* is responsible for storing the incoming packet in the memory of the FPGA, represented by the memory abstraction layer. For this, a memory address is requested from the *malloc* module. In the case of multiple external and internal memories, the allocated memory blocks are distributed equally over the available memory instances. The internal SRAM memory, if available, is only used for network packets up to a size of 512 bytes and is aligned in packet blocks of this size. In contrast to this, the external memory is 2048 byte aligned, sufficiently large for normal network packets with an MTU of 1514 bytes. As more than sufficient external DDR4 memory is available, the effect of wasting $\sim$ 500 bytes per packet is neglectable; however, it simplifies the address computation. If the internal SRAM memory for small packets is fully utilized, small packets can also be stored in external DDR4 memory with worse performance.

The *memory abstraction layer* forwards read and write requests to the corresponding memory instance based on the address of the request. Thus, only the *malloc* unit must be aware of the actual memory topology. The *rx-handler* and *tx-handler* can simply access all memory types in a unified way.

After an incoming packet is stored successfully in the memory and the abstraction layer confirms the successful write, the packet metadata information, i.e., the physical memory address of the packet, the length in bytes, and the queue ID, is handed over to the *queue memory*. The *queue memory* consists of a FIFO queue for each queue ID, and the incoming packet information is appended at the tail of the corresponding queue. If the queue reaches a configured maximum size, the newly arrived packet will be dropped, and the memory address is freed in the *malloc* unit, a typical taildrop behavior. The details of this module are discussed later.

On the other side of the *queue memory* in Fig. 3.17, the *scheduler* is attached. Depending on the scenario, different implementations can be instantiated here, all providing the same interface towards the *queue memory* and the *tx-handler*. The interface between the *queue memory* and the *scheduler* is used to find and dequeue packets and will be discussed later.

If the scheduler identifies a packet to be sent, the packet address and length are handed over to the *tx-handler*. Furthermore, a counter module is informed, counting the number of sent packets and bytes for each queue.

The *tx-handler* requests the packet data from the *memory abstraction layer* and sends out the data on the egress Ethernet port. This Ethernet port can be the same for ingressing packets, as the receiving and sending channels are independent of each other.

Non-blocking System Design

One main design goal of this prototype is high performance, i.e., high throughput, low jitter and a low base latency. Therefore, the task execution of the different logical modules should be decoupled as good as possible. This means, the beforehand introduced functional components should not wait for each other. e.g., the *rx-handler* must wait for the receive confirmation of metadata transmitted to the *queue memory*, currently being busy or locked by another task. To avoid this, we propose the concept of *job queues*. These queues can be instantiated between two modules with unidirectional communication and confirm the data reception immediately to the sending module. After being confirmed, the sending module can process the next packet, and the job queue waits until the receiver is ready to accept this data. In the overall design figure, they are only shown at the input of the *queue memory* and at the input of the *tx-handler*; however, they are instantiated between all modules, despite the interface from the *queue memory* and *scheduler*. The interface cannot be used in conjunction with a *job queue*, as the information flow is bidirectional. Job queues are also used for the *malloc* functionality, allocating memory addresses in advance.

Memory Access Parallelization

The utilized external DRAM memory technology, i.e., DDR4, suffers under high access latency and limited bandwidth. Note that the theoretical throughput is halved, as each packet must be written and read from the external memory. To increase the throughput, we have already introduced the concept of parallel memories and a round-robin arbitration between these memories at enqueue time. In addition to this, multiple overlapping read and write requests can be performed in parallel on the *memory abstraction layer*. We utilize read and write IDs of the AXI memory interface specification, allowing multiple unordered requests in parallel. The *rx-handler* is allowed to have up to four parallel outstanding write requests for each memory, e.g., 12 requests for two DDR4 memories and a single internal SRAM memory. This is mandatory, as we observed in our experiments a very high delay for write responses, i.e., the signal that a packet is successfully stored in the memory. Similarly, the *tx-handler* can address and prefetch multiple packets to be sent in parallel to utilize the memory bus as high as possible. Note that this optimization is currently optimized for the AMBA AXI memory interface, which is only available for Xilinx FPGAs. A migration to Intel FPGAs requires an adaption of the memory access or the loss of this improvement by parallelization.

Massive Parallel Packet Queueing

The queues within the FPGA are realized in the *queue memory*, holding a list of memory pointers on the packets in the external memory. This functionality is realized internally by four SRAM block memories, namely the *queues_mem*, *queue_valid_mem*, *buckets_mem*, and *bucket_valid_mem*.

In the *queue_valid_mem* memory, each bit indicates for one queue if at least one packet is in it, i.e., all non-empty queues have a 1 in this memory on the corresponding position. Each entry of this memory has a data width of 32 bits, which allows checking for 32 queues in parallel if a non-empty queue is available and can be suggested to the scheduler. In the given example of Fig. 3.19, the queues nr. 2 and 4 have at least one entry. If the first packet of a queue is pushed or the last packet popped, this bit is swapped.

The queues themselves are realized as a linked list. The *queues_mem* has a pointer to the first and last packet of each queue. In addition, the total amount of bytes in this queue is stored as a queue length, used for realizing the taildrop behavior. Every time a packet is pushed or popped, this value is updated incrementally.

Each packet is realized by an entry in the *buckets_mem*, pointing to the physical memory address of the packet. In addition, it holds the length of this packet and a pointer to the next packet in this queue, i.e., another *buckets_mem* entry. If this next pointer is $0xffffffff$, the tail of the queue is reached, and the last packet is popped. For performance reasons, the tail of a queue can be detected even simpler by

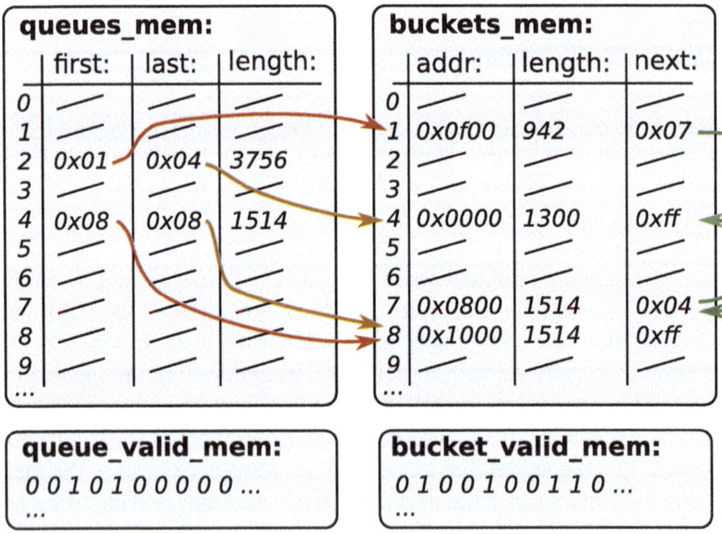

Fig. 3.19 Memory data structures of the queue memory storing packets of queues in a linked list. Each queue is identified by its first and last packet. A bucket holds the pointer on the packet in the external memory and to the next bucket in this list. Figure derived from: [24]. ©2022 IEEE. Reprinted, with permission, from "FPGA-assisted Massive Packet Queueing and Traffic Shaping at the Network Edge" [24]

comparing the first and last pointer of the *queues_mem* for equality. The next pointer is ignored in this case.

The last memory is the *bucket_valid_mem* and indicates the used entries in the *buckets_mem*. This memory is required to find free slots in the *buckets_mem* for new packets to be pushed, i.e., an allocation unit. Analogous, each dequeued packet must be freed in this memory. Note that the *buckets_mem* is a shared memory between all queues. Therefore the fixed resource utilization of an additional queue is only its first and last pointers in the *queues_mem* as well as its length register.

In the shown example of Fig. 3.19, queue nr. 2 holds three packets, and queue nr. 4 contains a single packet.

For enqueueing and dequeueing packets, two Finite-State Machines (FSMs) are realized in the *queue memory* module. We introduce a semaphore-based lock mechanism to avoid synchronous access to the memories by the enqueue and dequeue FSM. This mechanism prevents the two FSMs from operating on the same queue ID range, i.e., the same memory row in the *queue_valid_mem* at the same time. For example, no concurrent operations on all queues with the IDs 32 to 63 are allowed. By this, operations on the same *queues_mem* entry are prevented implicitly.

In the following, we describe the enqueueing mechanism in detail, as shown in Listing 3.2.

```
func enqueue(packet p, int qID):
  //check configured maximum queue length (taildrop)
  if (queue[qID].length > MAX_QUEUE_SIZE):
    //drop the packet
    external_mem_malloc.free(p.address)
    return

  //create buckets_mem entry
  bucket = new bucket()
  bucket.address = p.address
  bucket.length = p.length
  bucket.next = 0xffffffff

  //update queues_mem
  if (valid_memory[qID] == 0):
    //pushing the first packet
    queue[qID].first = bucket
  else:
    //update the currently last bucket entry of the queue
    queue[qID].last.next = bucket
  queue[qID].last = bucket
  queue[qID].length += p.length

  //mark queue to be non-empty (even if it was 1 before)
  valid_memory[qID] = 1
```

Listing 3.2 Working principle of the enqueue FSM of the queue memory

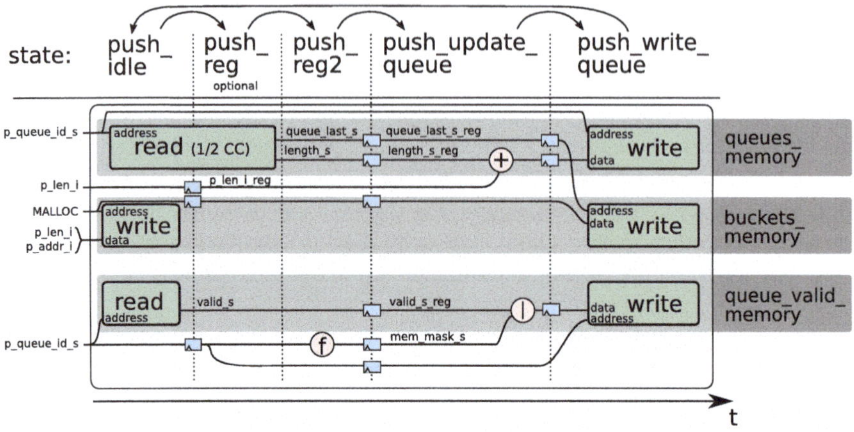

Fig. 3.20 FPGA internal finite state machine (FSM) and control signals of the push mechanism in the queue memory

First, the algorithm checks whether the current queue length has reached the taildrop limit (line 3). In this case, the packet is discarded, and the memory address of the packet is freed in the *malloc* unit. Otherwise, a new entry in the *buckets_mem* is created, pointing to the memory address of the packet (lines 9 to 12). The next pointer is $0xffffffff$, as newly pushed packets are always at the tail of the queue.

If the pushed packet is the first packet in the queue, i.e., the queue is currently empty, the *first* pointer must be redirected to the new *buckets_mem* entry (lines 15 to 17). In all cases, the *last* pointer will be updated to point to the newly added entry, and the queue length is increased by the length of the currently added packet. Otherwise, if the queue already has entries, the FSM must redirect the *next* pointer of the previous last bucket to the newly added entry (line 20). The bit in the *valid_memory*, indicating that the queue is not empty, can be set to 1 in all cases, even if it was 1 before (line 25), to avoid an unnecessary comparison.

This algorithm is described as a FSM in Verilog, consisting of 4 or 5 states (note one optional state), as shown in Fig. 3.20. In the figure, only the read and write operations on three memories are shown, as the *bucket_valid_mem* is not performance-critical and can be seen as an encapsulated allocation unit.

In the first clock cycle, during the state *push_idle*, the bucket entry is created and written. If the packet would be discarded later due to a taildrop, this memory is then simply freed. In parallel, a read operation on the *queues_memory* at the address of the queue ID is performed. This read operation is performed in either one or two clock cycles depending on the number of queues and FPGA clock frequency. In addition, the *queue_valid_memory* is read, as the queue where the packet is pushed might require an update of its valid bit. As this memory has a word size of 32 bits and only a complete word can be written, the remaining 31 bits must be known to update a single bit. The next state, *push_reg*, is only used if the number of queues and FPGA clock frequency is very high.

In the following two states, *push_reg2* and *push_update_queue*, the new values of these memories are computed, and the taildrop check is performed. Finally, in the state *push_write_queue*, the new memory data is written to the three memories. Note that each digital signal, e.g., *queue_last_s*, is stored after each state transition in a new data register to maximize the possible clock frequency of the FPGA. The rewriting for parallel execution of the presented algorithm is necessary to achieve the highest possible performance.

Assuming a clock frequency of 200 MHz, i.e., a clock cycle takes 5 ns, a theoretical enqueue packet rate of 40 million packets per second can be achieved. The dequeue state machine, following the same mechanisms, consists of 10 states, leading to a theoretical packet rate of 20 million packets per second. This state machine has more states as it (1) searches a non-empty queue, (2) requests a scheduler decision, and (3) dequeues the packet. However, this packet rate is still sufficient to achieve a line rate of 100 Gbit/s for large packet sizes and will be discussed later in evaluation Sect. 5.2.3.

Packet Scheduling at Scale

In the proposed architecture, the *scheduler* is attached to the *queue memory* on an interface, allowing to search non-empty queues and dequeueing them. In this work, we developed three scheduler implementations with different purposes:

- **No-Scheduler:** The intention of the *no-scheduler* is the performance evaluation of the remaining system components. It immediately accepts every suggested packet of the *queue memory* and is used for evaluation purposes only.
- **Simple-Scheduler:** The *simple scheduler* performs rate limiting for each queue based on a token bucket mechanism. It suits well for the demand in mobile access networks, i.e., as a part of the UPF.
- **HQoS-Scheduler:** The *HQoS-scheduler* fulfills the needs of residential access networks, mainly hierarchical rate limits and inter-subscriber QoS enforcement.

The interfaces of the three implementations towards the *queue memory* and *tx-handler* are identical and will be discussed in the following. Figure 3.21 depicts the simplified internal structure of the *HQoS-scheduler* and the interface towards the *queues_memory*. The *tx-handler* interface is not shown.

The scheduling process can be divided into four phases: (1) First, the *queues_memory* proposes a queue ID of a non-empty queue to the scheduler. Note that the packet length is not provided to the scheduler at this time. (2) Based on this queue ID, the scheduler decides if the first packet of this queue is allowed to be sent or not. For this, the internal token bucket instance(s) check if sufficient tokens are available, assuming maximum-sized packets, i.e., of the Maximum Transmission Unit (MTU). In the case of the hierarchical scheduler, multiple token bucket instances are instantiated in parallel, each representing a hierarchical layer. Only if all hierarchical layers allow a packet to be sent the scheduler output *ok_to_send* will be

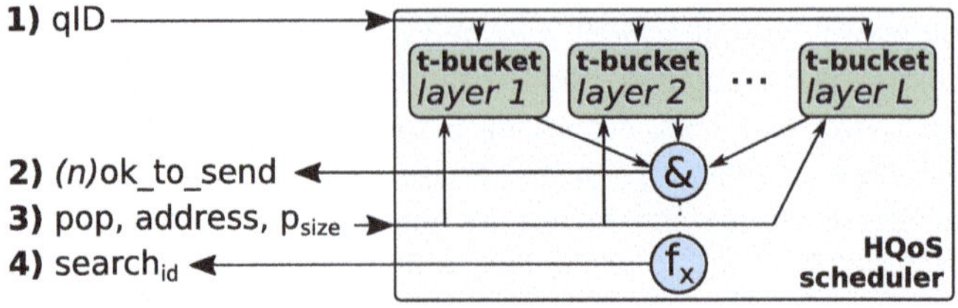

Fig. 3.21 Concept of the hierarchy-aware scheduler for residential access scenarios. The interface on the left side is connected to the *queue_memory*; the interface towards the *tx-handler* is not shown

set. (3) After the notification that a packet should be dequeued, the memory address and length of the packet are handed over from the *queues_memory* to the *scheduler*. In contrast to the scheduling decision, assuming a packet size of the MTU, the token buckets are updated with the actual packet length. This allows faster scheduling, as finding a non-empty queue involves only *queue_valid_memory*, while the packet length of the first packet is stored in the *buckets_mem*, and reading the length would require three subsequent memory accesses in total. Further, the packet address and length are handed over to the *tx-handler*. Last, in step (4), the scheduler notifies the *queues_memory* module to continue searching for non-empty queues. For this, the *scheduler* can provide a queue ID as starting index. If this index is the same as the currently popped packet, multiple packets of a single queue can be dequeued at once, assuming the queue is not empty. In the case of hierarchical access networks, the token bucket of a shared access node may prohibit sending a packet. Then, the *scheduler* can jump over all queue IDs belonging to this access node and continue searching for non-empty queues belonging to another access node. If a packet of the proposed queue ID should not be sent, i.e., the token bucket is empty, the *scheduler* can proceed with step 4 directly after the *queues_memory* offers the queue ID.

The token bucket module must maintain and update the states of many queues in parallel, e.g., for $\geq$100,000 rate-limited queues. A periodic update of the token bucket state, e.g., n tokens every 10 µs, would cause a significant overhead and is unnecessary as most subscribers do not permanently utilize their Internet connection.

Therefore, we propose a concept to update only the token buckets of active rate limiters. Every time a scheduler decision is requested, the following mechanism is used to update the token bucket value:

$$bucket[qID] = bucket[qID] + (t_{now} - t[qID]) \cdot token\_rate[qID]$$

$$t[qID] = t_{now}$$

For every queue, a *bucket* value, a timestamp (t), and a *token_rate* value are stored. The timestamp indicates when the bucket value for this queue ID was updated last. Based on this, the time difference to the current time (t_{now}) can be computed and multiplied by the number of tokens per time unit, i.e., the configured *token_rate*. This mechanism is realized within the FPGA, leveraging an 18-bit hardware multiplier, called DSP, to enable fast scheduling decisions. Timestamp overflows are detected by an overflow bit and can be handled. Note that an undetected overflow of the timestamp register can still occur, however, only if packets are in this queue. In this case, the overflow has no significant impact on the system behavior.

The scheduler implementation can be extended by an Active Queue Management (AQM) algorithm, discussed later in Sect. 3.3.3.

Control Plane Integration

Even though the focus of this work is on the data plane, considering the control plane integration is unavoidable. For each queue, the control plane can configure an individual rate. Further, the control plane can access configuration and status registers, e.g., taildrop queue depth or AQM support, and counter values. For this, a low performance data bus is available on the FPGA, allowing read and write requests via PCIe.

A simple control plane driver runs on the server connected to the FPGA via PCIe, converting incoming gRPC requests for read and write operations on the PCIe bus.

For this, two approaches exist, both implemented and evaluated in this work:

- **Kernel Module:** A custom kernel module is registered to be used for this PCIe device, i.e., by the vendor and device ID. This kernel module appears as a character device in Linux, e.g., as "*/dev/QosFpga*". In addition to the kernel module, a user space driver can access this character device and perform read and write operations on a custom API. This approach requires a kernel module compilation on every system and after every kernel update.
- **Memory Mapped I/O:** Instead of using a kernel module, the PCIe memory region of the FPGA can be memory mapped in the user space application, e.g., by mounting the following Linux path "*/sys/bus/pci/devices/0000:65:00.0/resource0*". This approach works without any kernel modification, only sufficient permissions are required. However, it is mandatory that neither interrupts nor Direct Memory Access (DMA) is used. As long as only configuration updates are performed, this is no disadvantage.

These two approaches are equivalent to the QoS co-processor functionality from the functional side. The memory mapping approach allows a simpler integration and maintenance, as no kernel module must be compiled and maintained. The control plane performance, i.e., how fast read or write operations can be performed, will be evaluated later in Sect. 5.2.3.

3.2.5 Discussion on Accelerated Internet Service Creation

In this chapter of the thesis, we presented approaches to accelerate residential and mobile Internet service creation. While the state-of-the-art approaches in current residential access networks are inflexible blackboxes, existing mobile access networks rely on softwarized network functions with limited performance. In this section, we proposed the concept of flexible hardware acceleration with good performance on P4-programmable switches and FPGAs as QoS co-processors for both access scenarios.

Our previously published results have already been picked up in related work, and other researchers have discussed similar topics in the meantime. In the following, we will discuss them briefly.

In 2019, Singh et al. presented an approach for offloading the virtual Evolved Packet Gateway (vEPG), the equivalent of the UPF in 4G networks, on P4 programmable switches [40]. Their work builds upon the concepts of the already published acceleration methods for the residential BNG use case, presented as part of this thesis. Further, their presented evaluation results are similar to the findings in this work, punctuating the results' validity of each other.

Nadas et al. presented a concept, named "Per Packet Value", enabling QoS enforcement in access networks [32]. This approach allows realizing prioritization between multiple subscribers in hierarchical residential access networks (compare Fig. 3.14). This approach would classify the packet at the access edge in the downstream direction and assigns a *per packet value*, which is later used in the access network to schedule packets. However, this approach would cause packet loss in the network elements of the access network, which is undesirable as one requirement is zero packet loss between the access edge and the subscriber.

Upon our ideas, the company APS Networks GmbH announced multiple products, building upon the conjunction of FPGAs and P4-programmable switching chips [3]. These switches are tailored to realize residential and mobile Internet service creation, but further use cases, e.g., deep packet inspection, are also intended.

The Telecom Infra Project, represented by multiple large Internet service providers, released in 2020 an open specification for the residential Internet service creation, named "OpenBNG" [41]. This work addresses the same needs and requirements as discussed in this work, i.e., functionality, flexibility, and performance.

Based on our results in this work, we suggest a further technological convergence of residential and mobile Internet access termination in the future, already discussed in the literature as Fixed Mobile Convergence (FMC) [8]. We extended our design, proposed in Sect. 3.3, allowing residential and mobile subscribers to be terminated simultaneously, both being authenticated by a 5G core.

In this design, all concepts of the preceding three sections are included. Note that in residential access networks, the control traffic is also sent to the access edge and will be redirected in the design to the slow path of the Access Gateway Function (AGF). Yet, this convergence was not investigated as part of this work.

Another related field of research investigates the combination of software- and hardware-based data planes. This idea of a hybrid network function, consisting of hardware and software parts for packet processing, was discussed in the literature, e.g., by Katta et al. [18]. These approaches are compatible with the presented concepts in this work and allow future space for improvement, i.e., distributing the total set of subscribers between programmable hardware and a software switch. By this,

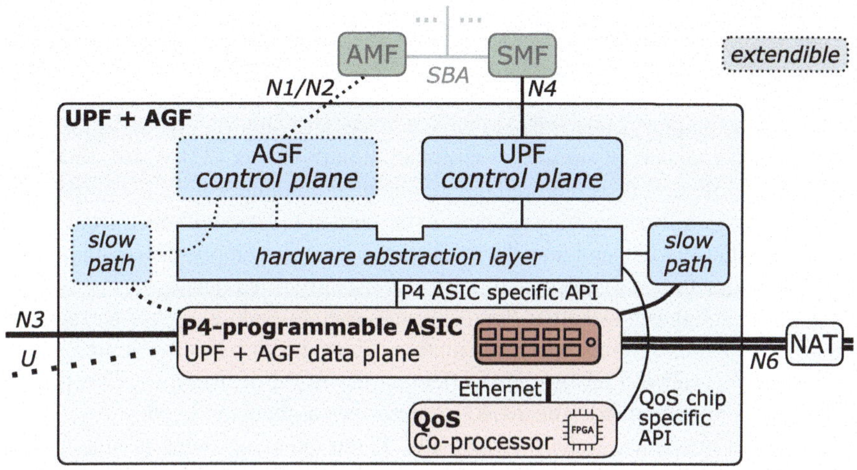

Fig. 3.22 Concept of a general UPF design for fixed-mobile convergence following the concepts of this work and the TR-470 specification [8], realizing residential and mobile subscriber termination in the same dataplane

the almost infinite memory capacity of software switches is combined with high-performance hardware. In the presented design for residential and mobile subscriber termination in Fig. 3.22, the slow path represents a software switch that could be extended following the idea of a hybrid switch. By this, the number of parallel subscriber sessions could be further increased; however, this is currently not needed.

3.3 Active Queue Management

One goal of this work is the QoS improvement by hardware acceleration in networking edge environments. In the previous sections, we presented concepts for Packet I/O in hardware accelerators with low latency and high throughput; further, we discussed approaches to realize residential and mobile Internet service creation on programmable hardware. However, the previously presented concepts aim only at a high QoS of the network function but not of the end-to-end connection, e.g., from the subscriber to the Internet service.

As introduced in the background Sect. 2.4, traffic shaping of congestion-controlled network flows, e.g., common TCP flows in the Internet, tend to fill the shaper queue completely, a phenomenon called "bufferbloat". Active Queue Management (AQM) algorithms offer a remedy to this. In this section, we introduce concepts how existing algorithms, tailored for software-based network stacks, can be adapted for and integrated in programmable hardware.

3.3.1 The CoDel Algorithm

In the following, we analyze the *Controlled Delay (CoDel)* algorithm, which we choose as an exemplary mechanism to avoid unnecessarily high delays in the Internet, extending the concepts of subscriber termination in this work [33]. Traffic shaping, i.e., queueing of packets, is required before potential bottleneck links to compensate for the rate variability of arriving packets. In the case of bloated buffers, the queueing delay is constantly high, and thus the queue never becomes empty. The main idea of the CoDel algorithm is to detect and prevent constantly filled queues.

In Listing 3.3, the CoDel algorithm is given as pseudo-code, extracted and simplified from the Linux reference implementation. The algorithm has two input parameters, *TARGET* and *INTERVAL*, both affecting the AQM decision, i.e., if a packet should be dropped/marked. The *TARGET* describes a queueing latency, e.g., 5 ms, which should not be exceeded permanently. If the queueing delay is higher than the *TARGET* value for a time period longer than *INTERVAL*, the algorithm starts dropping or marking packets. The default *INTERVAL* value of 100 ms is sufficient to absorb temporary packet bursts.

Next, we analyze the CoDel algorithm in detail. In general, CoDel has a state, either being *idle* or *dropping*. The algorithm is called for each packet to be scheduled, compare Line 5 of Listing 3.3. First, in Line 7, the current queue delay is checked (*if_1*). If the current queue delay is below the latency *TARGET* or less than one MTU-sized packet is in the queue, the algorithm transits to the *idle* state. The MTU check is only relevant for very low shaper rates to ensure that the algorithm can send at least one MTU-sized packet. In this case, the algorithm terminates without dropping any packet (Line 9).

Otherwise, the algorithm checks if the current packet is the first one exceeding the *TARGET* delay (*if_2* in Line 11). In this case, the *dropping* state is entered, and the number of dropped packets in the previous dropping phase is saved (Line 13 and 22). In Line 21, the algorithm computes the point in time when the first packet should be dropped. Note that this calculation includes a square root computation and a division, which is complex to implement in hardware.

In Line 15, the basic algorithm is extended by an optimization (*if_3*). In the case of an increasing number of dropped packets over the last two dropping phases, the algorithm starts with a *count* value bigger than 1 (Line 18), leading to earlier and more frequent packet drops. This is a very simple learning mechanism, which allows reacting more effectively under load-intensive network conditions, e.g., many parallel and bandwidth-demanding TCP flows.

The actual packet dropping is performed in Line 26 and following (*if_4*). If the algorithm remains in the *dropping* state until the time *drop_next_packet*, a packet is dropped, the dropped packet counter is increased, and the next time to drop a packet is computed. Note that if the queueing latency falls below the *TARGET* latency, the state toggles and the timestamp of the next packet to be dropped is not used anymore.

```
#define TARGET 5 //ms
#define INTERVAL 100 //ms
Queue queue; State state

for each received packet p:
 //Is the latency below the target delay? (if_1)
 if(p.queue_delay < TARGET || queue.byte < IFACE_MTU):
   state.dropping = false  //false: idle, true: dropping
   continue
 //Do we have to enter the drop state? (if_2)
 if(state.dropping == false):
   state.dropping = true
   tmp_count = state.count
   //Was the number of dropped packets per round increasing?(if_3)
   if((state.count - state.last_count > 1) &&
       (now - s.drop_next_packet < 16*INTERVAL)):
     //Start packet dropping earlier
     state.count = state.count - state.last_count
   else:
     state.count = 1
   state.drop_next_packet = now + INTERVAL / sqrt(state.count)
   state.last_count = tmp_count
   continue
 //Are we in droping state and reached the time to drop the next packet?
 //(if_4)
 if(state.dropping && state.drop_next_packet <= now):
   p.drop()
   state.count++
   state.drop_next_packet = now + INTERVAL / sqrt(state.count)
```

Listing 3.3 Pseudo-code for the CoDel AQM algorithm, based on RFC 8289 [33] and [25]

3.3.2 P4-Codel

In this section, we investigate how the CoDel algorithm, introduced in Sect. 3.3.1, can be adapted for P4-programmable hardware. First, we investigate the reduction of the mathematical complexity. Second, we propose a concept for mapping the algorithm in a cycle-free way on the programmable hardware resources. The underlying source code of this work is available open-source.[2]

[2] https://github.com/ralfkundel/p4-codel.

Reduction of Mathematical Complexity

In Line 21 of Listing 3.3, the CoDel algorithm includes an inverse square root computation. However, even highly specialized hardware architectures for (inverse) square root computation require many clock cycles, e.g., 12 clock cycles as named by Hasnat et al. in 2017 [13]. Existing P4-programmable hardware would not allow the realization of such calculations, and a realization in FPGAs would delay scheduler decisions. Therefore, to ensure a constant and low processing time in hardware pipelines, this function should be approximated.

Although the mathematical approximation of a function causes an approximation error, this error is neglectable in this scenario for two reasons: First, the computed value is not used as the input for further direction, which would cause the error to be multiplied by itself and by that increased. Second, the congestion control mechanisms and flow mixture in the Internet depend on many influencing factors, causing a non-deterministic behavior. Therefore, any AQM algorithm is only an estimation to handle this.

Therefore, we propose a simple approximation for the squareroot function, which can be realized with ternary match tables of P4-programmable switches and with logical building blocks of FPGAs. The input of this function is the integer number n. The approximation performs a ternary lookup on the number of leading zeros of the binary input n, and the lookup table entry contains a precomputed value for the function. This mechanism can be realized with existing hardware concepts. Figure 3.23 depicts the original inverse square root function and the approximation. This function could be approximated quite well by a very low number of entries. This approximation can be improved by utilizing range matches instead of counting the number of leading zeros; however, this is not needed for this algorithm.

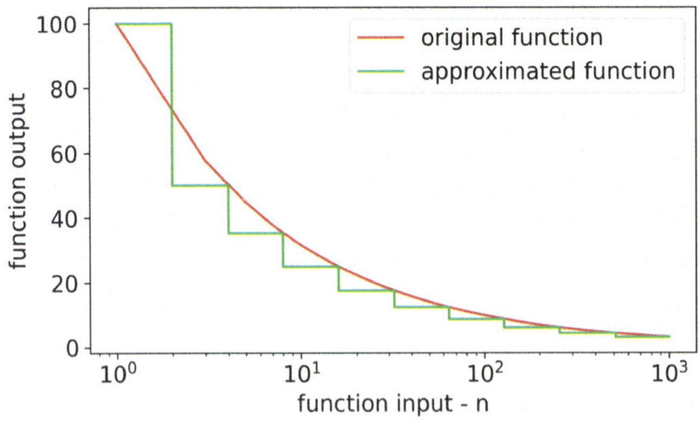

Fig. 3.23 Mathematical approximation of the $\frac{INTERVAL}{\sqrt{n}}$ function to realize to CoDel AQM in programmable hardware. The approximation performs a ternary lookup of the input n, i.e., counting the number of leading zeros in binary representation

To summarize, we introduced an approximation for the inverse square root function and have shown its mathematical matching with the original function. However, this assessment does not allow any statement about the real influence of the approximation on the end-to-end behavior of network flows and congestion. Later in Sect. 5.3, we investigate how the approximated CoDel algorithm behaves compared to the Linux reference implementation.

Pipeline Mapping on P4-Programmable Switches

The next step in realizing the CoDel algorithm in P4-programmable switches is the rewriting of the algorithm, allowing the compiler to map the functionality on the available switch resources.

The general concept of P4 switches, introduced in Sect. 2.3, relies on a packet processing pipeline, as shown in Fig. 3.24. Based on the assumption that the packet queues are located between the ingress and egress pipeline, the per-packet queueing delay is only available in the egress pipeline. Therefore, the CoDel algorithm should be instantiated in the egress pipeline. Note that asynchronous feedback of the current queue occupancy in the ingress pipeline is theoretically possible and might be supported by future generations of P4-switches. The CoDel AQM algorithm could also be instantiated before the queues if asynchronous feedback is supported. However, this is not the case for current generations of P4 switches and, thus, we expect the algorithm to be located behind the queues.

Next, we transform the CoDel algorithm, introduced in Listing 3.3, into a state-centric dependency graph, shown in Fig. 3.25. This graph highlights the three stateful variables, *dropping*, *count*, and *last_count*. It is visually recognizable in the dependency graph that there is a time gap between reading operations, denoted with an **r**, and writing operations, denoted with a **w**. However, each stateful variable can and must be placed only in a single pipeline stage of the P4-switch. Cyclic dependencies cannot be supported by P4-programmable switches, as this would cause a data inconsistency between multiple packets and an information flow in the pipeline contrary to the pipeline flow.

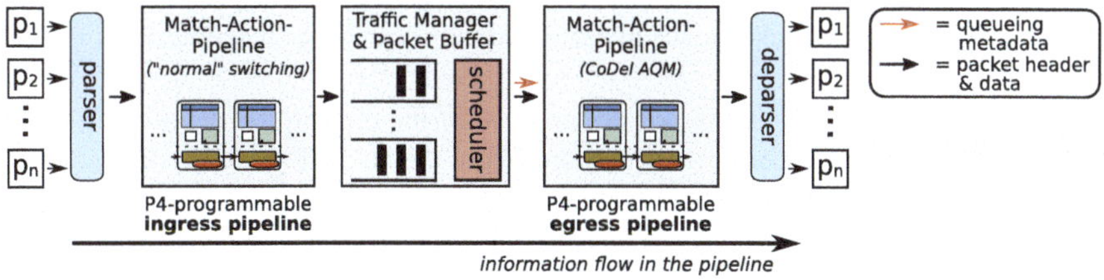

Fig. 3.24 Structure of the P4-programmable hardware pipeline. Packet data can flow only from left to right, the queueing latency of a packet is only available in the egress pipeline. ©2021 IEEE. Reprinted, with permission, from "P4-CoDel: Experiences on Programmable Data Plane Hardware" [25]

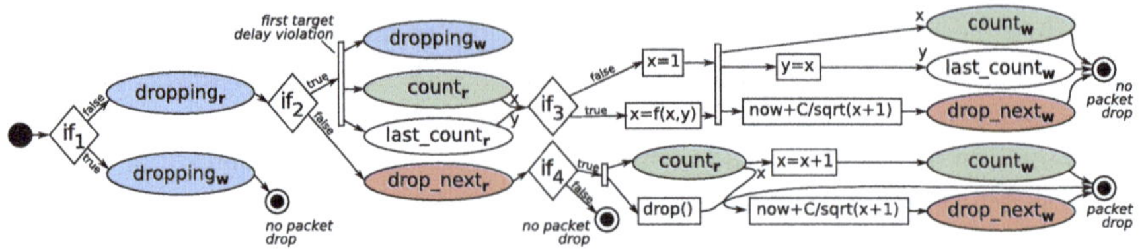

Fig. 3.25 State-centric dependency graph of the CoDel algorithm, as specified in its Linux reference implementation [33]. Memory operations are denoted as read (r) or write (w). ©2021 IEEE. Reprinted, with permission, from "P4-CoDel: Experiences on Programmable Data Plane Hardware" [25]

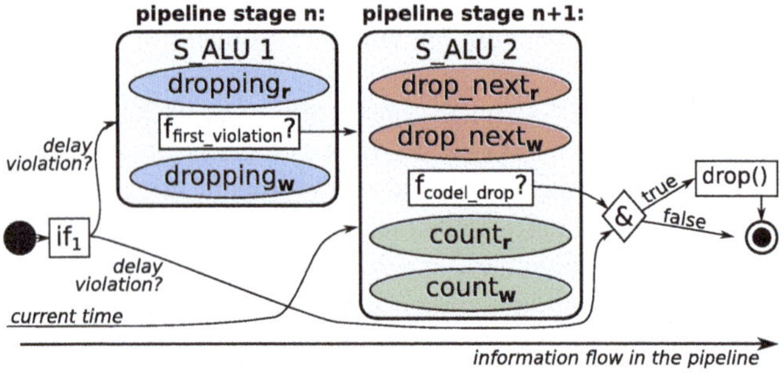

Fig. 3.26 Cycle-free dependency graph placement of the CoDel algorithm, optimized for the hardware resources of the P4-programmable Intel Tofino. The grey boxes represent stateful ALUs of the hardware, a component that allows atomic and stateful operations for all network packets. ©2021 IEEE. Reprinted, with permission, from "P4-CoDel: Experiences on Programmable Data Plane Hardware" [25]

Therefore, each stateful variable must be realized within a register located in a single pipeline stage of the P4 switch. All computational logic between reading and writing this register must also be located in the same pipeline stage.

For example, the *dropping* variable is read to determine the current state. If this state is *idle* and the *TARGET* queue delay is violated, the *dropping* state is entered. This state entering should be detected for further computations. The read and write operations on the *dropping* register must be located in the same pipeline stage. Otherwise, the subsequent packet reads an outdated value, and the state entering may be detected twice.

Listing 3.4 shows a cycle-free representation of the algorithm optimized for the P4-programmable Intel Tofino architecture. The sketch in Fig. 3.26 shows logic mapping on the corresponding pipeline stages.

First, for each packet, a possible *TARGET* delay violation is examined by if_1. This function does not depend on any state and can be performed on the packet metadata.

The following stateful operations are executed for all packets, independent of a delay violation. In the case of no delay violation, the final logical *AND (&)* operation disables a possible dropping decision, and the stateful CoDel logic has no effect.

This optimization simplifies the behavior of the two stateful ALU operations in the pipeline stage n and n+1, and enables the logic to be compiled.

```
#define TARGET 5 //ms
#define INTERVAL 100 //ms
Queue queue;

for each received packet p:
  //check for target delay violation? (if_1)
  if(p.queue_delay < TARGET || queue.byte < IFACE_MTU):
    delay_violation = false
  else:
    delay_violation = true
  is_first_violation = S_ALU1.exec(delay_violation)
  codel_drop = S_ALU2.exec(is_first_violation)
  if(delay_violation && codel_drop): //&
    p.drop()
```

Listing 3.4 Cycle-free representation of the CoDel algorithm optimized for P4-programmable pipelines. Code derived from [25]

The state management of the CoDel algorithm is located in the first stateful ALU (*S_ALU1*). The input of this component is the information on whether there is a *TARGET* delay violation or not. The output indicates if the state has changed from *idle* to *dropping*, i.e., the current packet is the first packet with a delay violation.

The input of the second stateful ALU (*S_ALU2*) is the information on whether it is the first queue delay violation or not. Further, the *current time*, an incremental integer number, is the second input. The information on the algorithm state, i.e., if it is in the *dropping* state or not, is not known. This ALU maintains two stateful registers, *drop_next* and *count*, both having a width of 32 bit. If the current packet is the first delay violation, the *count* register will be set to 1. In addition, the time *drop_next*, describing the first packet to be dropped, is computed.

If the current packet is not the first delay violation, a comparison between the current time and the *drop_next* register is performed. As soon as the current time overruns this value, the output of this ALU is set to *codel_drop = 1*, the *counter* is incremented, and the next *drop_next* value is computed.

As the number of inputs and registers per stateful ALU is limited to two by currently available P4-hardware, the *last_count* register and corresponding optimization in Line 15 of Listing 3.3 cannot be realized. If future P4-programmable hardware supports more stateful registers per pipeline stage and more input signals, i.e., the *dropping* state, an improved version of CoDel could be realized.

Pipeline Mapping on P4-Programmable Network Interface Cards

Besides the Intel Tofino platform, we briefly discuss Netronome SmartNICs as a second P4-programmable target for queue management. This P4 target is already known from Sect. 3.2, where we discussed Internet service creation. This platform consists of a Network Processing Unit (NPU), which executes assembly code provided by a P4 or C compiler. The execution is distributed over multiple cores, processing packets in parallel. To the best of our knowledge, details on the compilation and mapping process of P4-programs on this hardware architecture are not known.

While investigating the capabilities of this platform, we observed several issues, probably caused by limited compiler functionality. First, the current queue delay is not accessible by the P4 program. To circumvent this issue, we utilized embedded C functions in the C code, which access the corresponding information. As the information is available, we assume this to be a compiler limitation only.

Further, this data cannot be used as an input for further processing within the P4 program. Therefore, we implement the complete CoDel AQM algorithm as an external C function, analogous to the pseudo-code in Listing 3.3. Note that the complex square root operation is still approximated by counting the number of leading zeros. The observed limitation of programmable data plane hardware regarding complex mathematical operations is generous and is not caused by any programming language.

We do not have detailed information on the parallel packet processing within the SmartNIC. However, it is very likely that the parallel execution of the same program on multiple processor cores will cause race conditions on the shared register values of the algorithm. Later, in Sect. 5.3, we evaluate this solution and compare it with the Linux reference implementation. Assuming the register values are not synchronized and thread-safe between the processing cores, as implied in the hardware datasheets, we expect a higher packet drop rate due to simultaneous dropping decisions.

3.3.3 FPGA-Based Queue Level Control

In Sect. 3.2.4, we introduced a concept for realizing QoS functionality on FPGAs, aiming at per-subscriber traffic shaping and hierarchical scheduling in Internet access networks. This concept, however, would also benefit from AQM, actively avoiding *bufferbloat* to improve the end-to-end service quality and experience. In this section, we discuss how the CoDel algorithm can be integrated in this previously presented design and propose a prototypical implementation. Analogous to the restrictions of P4-programmable data planes, a software algorithm cannot be instantiated as it is on FPGAs, but it must be adapted.

Modular Integration and Interaction of the AQM Algorithm in the Scheduler

The previously introduced FPGA architecture contains an exchangeable *scheduler* module (compare Fig. 3.17). Within this module, the rate-limiting logic should be extended by the CoDel AQM functionality. For this, we extend the general state machine of the scheduler, currently only requesting feedback from one or multiple rate limiters, by an AQM instance. The general control flow of this state machine is presented in Listing 3.5.

The scheduler is called for each non-empty queue, suggested by the *queue memory*. First, in Line 4, the scheduler iterates over all rate limiter instances and checks if a packet of the given queue is allowed to be sent. In the case of mobile access networks, typically, only a single rate limiter exists. For hierarchical residential access networks, multiple parallel limiters are instantiated. Note that the sequential check of multiple rate limiters is realized by a logical *AND* operation in hardware, i.e., all rate limiters are requested for their packet acceptance (*ok*) in parallel (compare Sect. 3.2.4). If at least one scheduler declines the packet to be scheduled (Line 6), the packet is not sent now, and the process terminates. Otherwise, the logic proceeds with requesting an AQM decision.

The input of the AQM module, called in Line 11, is the queue ID, the rate limit of the queue, and the current queue length. While the *queue memory* provides the queue ID and length, the queue rate must be provided from the configuration memory of the per-subscriber rate limiter.

```
1  for each suggested queue q:
2    ok = true
3
4    for each limiter in rate_limiters:
5      ok = ok && limiter.is_ok(q.id)
6    if(!ok):
7      //at least one rate limiter declined packet sending
8      continue
9
10  //AQM integration:
11  aqm_decision = aqm.request(q.id, q.rate, q.length)
12  if (aqm_decision == drop):
13    //the first packet in the queue is dropped for AQM reasons
14    q.pop_and_drop()
15
16  // if the scheduler accepts the offered queue, always a packet is
      sent
17  q.pop_and_send()
```

Listing 3.5 Integration of the AQM functionality in the scheduler module in extension to the FPGA design presented in Fig. 3.21

The queue rate and length are required, as the latency of the packet is computed by dividing the queue length by the queue rate. Alternatively, the enqueue timestamp must be stored for each packet to compute the queue delay. Both approaches are viable; however, the chosen method requires less memory as no timestamp must be stored. On the other hand, storing a timestamp for each packet simplifies the queue delay computation, as discussed in the following subsection.

If the AQM instance decides that a packet should be dropped, this packet is popped from the queue memory and then discarded instead of handing it over to the *tx-handler* (Line 14). Further, in Line 17, a second packet will be popped and sent from the same queue. This is a viable decision, as the goal of the AQM is not to reduce the scheduling rate; instead, the congestion control mechanism should be notified by controlled packet drops. In the case of no AQM drop decision, only this *pop_and_send* logic is executed.

Note that the scheduling logic is presented and described as a sequential process. However, the actual FPGA realization is highly parallelized to achieve faster scheduling decisions.

FPGA-Based CoDel Realization

After discussing the overall integration of the CoDel AQM logic in the scheduler module, we face the realization of the algorithm next. The block diagram in Fig. 3.27 shows the internal logical structure. It consists of three data input signals, *queue_id*, *queue_rate*, *queue_length*, and a *request* indication signal.

If a request is performed, first, the queueing delay is computed by dividing the *queue_length* by the *queue_rate*. Due to its complexity, this operation requires two clock cycles within the FPGA and is already optimized by allowing a little inaccuracy, i.e., the *queue_length* is rounded to multiples of 64 byte. As stated before, this could also be realized by storing the enqueue timestamp for each packet and subtracting it from the current time. A subtraction is significantly easier than a division. However, storing additional per-packet metadata, i.e., the enqueue timestamp, requires additional memory resources, lowering the total number of queues that can be realized. Next, a possible *TARGET* delay violation is checked. Finally, the CoDel AQM logic is realized as a Finite-State Machine (FSM). Besides the already discussed inputs, a local clock is instantiated.

For each queue ID, there is an entry in the *queue_state_mem*, responsible for maintaining the CoDel state, as discussed in Sect. 3.3.2. Specifically, a 1-bit state (*dropping* or *idle*), and a 31-bit timestamp are stored. The timestamp indicates the *drop_next_packet* value, already introduced in the P4-realization section. The logic of the FSM is realized analogous to the P4 switch implementation.

Note that at any time only a single decision request is processed within this AQM module and the whole scheduler.

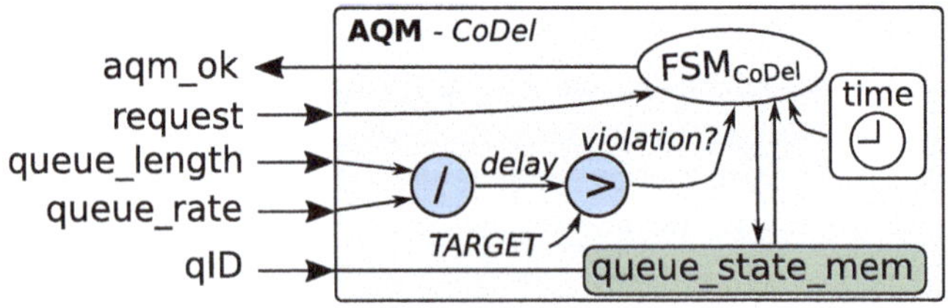

Fig. 3.27 Internal structure of the CoDel Active Queue Management (AQM) module, realized as a low-level FPGA module

3.3.4 Discussion on AQM in Programmable Hardware

In this thesis chapter, we introduced concepts to realize Active Queue Management (AQM) algorithms in programmable data planes, specifically, P4-programmable switches and FPGAs. The previously introduced concepts for Internet service creation, including subscriber termination and QoS-aware rate enforcement, were extended. This allows an improved end-to-end QoS in Internet access networks. Specifically, the deployed AQM algorithm allows a lower queueing delay while the same throughput can be achieved. New mechanisms in the data plane are also conceivable [22].

Other researchers investigated similar topics in parallel or building upon our work. Extending our initial work on the CoDel algorithm realization in the P4 programming language [19], Papagianni et al. proposed a novel algorithm, "PI2", tailored for P4-programmable hardware [37]. However, they addressed only a realization in the P4 software switch *bmv2* that does not suffer under the hardware restrictions discussed in this section. Therefore, it is not clear if their algorithm is a viable solution for hardware switches. However, it emphasizes the expressiveness of the P4 programming language and the concept of AQM realizations in hardware.

Harkous et al. presented a virtual queue-based traffic management approach in P4 data planes, including an AQM-based Explicit Congestion Notification (ECN) mechanism [12]. They faced the AQM mechanism of the newly introduced Low Latency, Low Loss, Scalable Throughput (L4S) congestion control scheme [5]. The realized prototype is based on the P4 software switch *bmv2* and P4-programmable smart network interface cards from Netronome, the same as utilized in this work. The authors have shown that different QoS classes, named *slices*, can be separated by their approach, and different service levels can be guaranteed. In their realization with the P4-programmable hardware from Netronome, they could not utilize the queueing capabilities of the hardware, as "some of the standard metadata fields that report the queue occupancy are missing" [12]. This description coincides with the findings in this work. However, in our approach, we circumvented this limitation by addressing the queue level status register by embedding micro-C code in the P4 program, an unclean solution but viable to prove the general feasibility.

Kunze et al. investigated the powerfulness of the Intel Tofino platform in conjunction with the P4 programming language for AQM in general [26]. The authors state that the realization of queue management algorithms in P4 hardware is very challenging, and existing software algorithms cannot be used without modifications. This statement agrees with the findings in this work and partially builds upon our previously published research results. Further, they demand better support for queue management in the P4 programming language. We consider this critical, as it is not compatible with the basic principles of a cycle-free switch pipeline. We expect a hybrid chip design of future programmable switches as a viable solution, consisting of P4-programmable pipelines, fixed data structures for packet queues, and FPGA-like programmable regions to describe custom schedulers, including cyclic algorithms. Programmable packet processors, such as the used *P4-SmartNICs* in this work, may also be appropriate programmable hardware for packet schedulers.

References

1. 3GPP. GPRS Tunnelling Protocol (GTP) across the Gn and Gp interface. Technical report (TR). Version 15.5.0. June 2019
2. 3GPP. Release 15 Description; Summary of Rel-15 Work Items. Technical report (TR). Version 15.0.0. Oct. 2019
3. APS Networks GmbH. APS2172Q (Tofino+FPGA). https://www.aps-networks. com/products/aps2172q/. [Online; Accessed 06-May-2022]
4. Bittner R, Ruf E (2014) Forin A (2014) Direct GPU/FPGA communication Via PCI express. Cluster Comput 17(2):339–348
5. Briscoe B, De Schepper K, Bagnulo M, White G (2022) Low latency, low loss, scalable throughput (L4S) internet service: architecture. In: InternetDraft. Work in progress. Internet engineering task force, p 45. https://datatracker.ietf.org/doc/html/draft-ietf-tsvwg-l4sarch-17
6. Bronstein Z, Roch E, Xia J (2015) Molkho A (2015) Uniform handling and abstraction of NFV hardware accelerators. IEEE Netw 29(3):22–29
7. Carrel D, Evarts J, Lidl K, Louis A. Mamakos, Simone D, Wheeler R (1999) A method for transmitting PPP Over Ethernet (PPPoE). RFC 2516. Feb. 1999. https://www.rfc-editor.org/info/rfc2516
8. Christele B, Paul M (2022) 5G wireless wireline convergence architecture. Technical report TR-470. Broadband Forum, Mar 2022
9. Developing a Linux Kernel Module using GPUDirect RDMA. Technical Documentation. [Online; Accessed 27-February-2022]. 2022. https://docs.nvidia.com/cuda/gpudirect-rdma/index.html
10. The Linux Foundation (2010). https://www.dpdk.org/
11. free5gc (2019). https://github.com/free5gc/free5gc
12. Harkous H, Papagianni C, De Schepper Koen, Jarschel M, Dimolianis M, Pries R (2021) Virtual queues for P4: a poor man's programmable traffic manager. IEEE Trans Netw Serv Manag 18(3):2860–2872
13. Hasnat A, Bhattacharyya T, Dey A, Halder S, Bhattacharjee D (2017) A fast FPGA based architecture for computation of square root and inverse square root. In: 2017 Devices for Integrated Circuit (DevIC), pp 383–387
14. Hastings N, McLean P (1996) TCP/IP spoofing fundamentals. In: Conference proceedings of the 1996 IEEE fifteenth annual international phoenix conference on computers and communications, pp 218–224
15. Host Bypassing (2021). https://github.com/ralfkundel/HostBypassing
16. IEEE Standard for Verilog Hardware Description Language. In: IEEE Std 1364-2005 (Revision of IEEE Std 1364-2001) (2006), pp 1–590
17. Intel Corporation (2012) Intel 82599 10 GbE Controller. Datasheet
18. Katta N, Alipourfard O, Rexford J, Walker D (2016) Cacheflow: dependency-aware rule-caching for software-defined networks. In: Proceedings of the symposium on SDN research, pp 1–12
19. Kundel R, Blendin J, Viernickel T, Koldehofe B, Steinmetz R (2018) P4-CoDel: active queue management in programmable data planes. In: Proceedings of the IEEE conference on network function virtualization and software defined networks (NFV-SDN). IEEE, pp 1–4
20. Kundel R, Burkert T, Griwodz C, Koldehofe B (2019) Chaining of hardware accelerated virtual network functions in PCIe environments. In: Proceedings of the 20th international middleware conference demos and posters. Middleware '19. ACM, pp 13–14
21. Kundel R, Eryigit K, Markussen J, Griwodz C, Abboud O, Hark R, Steinmetz R (2021) Host Bypassing: direct data piping from the network to the hardware accelerator. In: Proceedings of the 14th international symposium on embedded Multicore/Many-core Systems-on-Chip (MCSoC). Best Paper Award. IEEE, pp 23–30
22. Kundel R, Krishna NB, Gärtner C, Meuser T, Rizk A (2021) Poster: reverse-path congestion notification: accelerating the congestion control feedback loop. In: Proceedings of the 29th international conference on network protocols (ICNP). IEEE, pp 1–2

23. Kundel R, Nobach L, Blendin J, Maas W, Zimber A, Kolbe H-J, Schyguda G, Gurevich V, Hark R, Koldehofe B, Steinmetz R (2021) OpenBNG: Central office network functions on programmable data plane hardware. Int J Netw Manag 31(1):1–25
24. Kundel R, Nobach L, Kolbe H-J, Meuser T, Steinmetz R (2022) FPGA-assisted massive packet queueing and traffic shaping at the network edge. In: Proceedings of the 30th annual international symposium on field-programmable custom computing machines (FCCM). IEEE, p 1
25. Kundel R, Rizk A, Blendin J, Koldehofe B, Hark R, Steinmetz R (2021) P4-CoDel: experiences on programmable data plane hardware. In: Proceedings of the IEEE international conference on communications (ICC). IEEE, pp 1–6
26. Ike K, Moritz G, David S, Klaus W, Jan Rüth (2021) Tofino + P4: a strong compound for AQM on high-speed networks? In. IFIP/IEEE international symposium on integrated network management (IM) 2021:72–80
27. Kurauchi Y (2019) go-pfcp. https://github.com/wmnsk/go-pfcp
28. Li S, Reddy D, Jacob B (2018) A performance and power comparison of modern high-speed DRAM architectures. In: Proceedings of the international symposium on memory systems. MEMSYS '18. ACM, New York, pp 341–353
29. Markussen J, Kristiansen LB, Borgli RJ, Stensland HK, Seifert F, Riegler M, Griwodz C, Halvorsen P (2020) Flexible device compositions and dynamic resource sharing in PCIe interconnected clusters using device lending. Cluster Comput 23:1211–1234
30. Markussen J, Kristiansen LB, Halvorsen P, KiellandGyrud H, Stensland H, Griwodz C (2021) SmartIO: zero-overhead device sharing through PCIe networking. ACM Trans Comput Syst 38(1–2):2:1–2:78
31. Markussen J, Kristiansen LB, Stensland HK, Seifert F, Griwodz C, Halvorsen P (2018) Flexible device sharing in PCIe clusters using device lending. In: Proceedings of the 47th international conference on parallel processing companion. ICPP '18. ACM, pp 48:1–48:10
32. Nadas S, Turanyi ZR, Racz S (2016) Per packet value: a practical concept for network resource sharing. In: 2016 IEEE global communications conference (GLOBECOM, 2016, pp 1–7
33. Nichols K, Jacobson V, McGregor A, Iyengar J (2018) Controlled delay active queue management. RFC 8289. Jan 2018. https://doi.org/10.17487/RFC8289. https://www.rfc-editor.org/info/rfc8289
34. Nobach L, Blendin J, Kolbe H-J, Schyguda G, Hausheer D (2017) Bare-metal switches and their customization and usability in a carrier-grade environment. In: 2017 IEEE 42nd conference on local computer networks (LCN), 2017, pp 649–657
35. Leonhard N, David H (2015) Open, elastic provisioning of hardware acceleration in nfv environments. In: 2015 international conference and workshops on networked systems (NetSys). IEEE, pp 1–5
36. NVIDIA Corporation (2022) CUDA toolkit documentation v11.6.0. https://docs.nvidia.com/cuda/. [Online; Accessed 07-February-2022]. Jan 2022
37. Papagianni C, De Schepper K (2019) PI2 for P4: an active queue management scheme for programmable data planes. In: Proceedings of the 15th international conference on emerging networking experiments and technologies. ACM, pp 84–86
38. PCI Express 3.1 Base specification. Peripheral component interconnect special interest group (PCI-SIG). 2010
39. Perino D, Varvello M (2011) A reality check for content centric networking. In: Proceedings of the ACM SIGCOMM workshop on Information-centric networking, pp 44–49
40. Singh SK, Rothenberg CE, Patra G, Pongracz G (2019) Offloading virtual evolved packet gateway user plane functions to a programmable ASIC. In: Proceedings of the 1st ACM CoNEXT workshop on emerging in-network computing paradigms. ENCP '19. ACM, pp 9–14
41. Telecom Infra Project (TIP) (2020) Open BNG - Technical Requirements. Technical report
42. Thoma Y, Dassatti A, Molla D (2013) FPGA2: An open source framework for FPGA-GPU PCIe communication. In: 2013 international conference on reconfigurable computing and FPGAs (ReConFig), pp 1–6

Chapter 4
P4STA: High Precision Network Function Benchmarking

Network function benchmarking is crucial for system understanding and optimization in computer networks, including the approaches presented in the prior Chap. 3 of this work. Specifically, an accurate capturing of the Quality of Service (QoS) identifiers latency, packet loss, throughput, and jitter is fundamental.

These metrics indicate if the tested network function fulfills the postulated service quality. Further, they can be used to understand and optimize network functions. Besides latency, packet loss is a very important metric: Even slight packet loss, especially at the end of data flows, can have a strong negative influence on the flow completion time [8]. In addition, Internet access networks must guarantee zero undetected packet loss for regulatory reasons (compare Sect. 2.2). Therefore, network function benchmarking is very important to achieve an improved network quality.

The importance and difficulty of network function benchmarking become apparent when we examine current networking hardware, which experienced many performance jumps in the past: An up-to-date data center switch may consist of 64 ports, each operating at 100 Gbit/s. At this link speed, transmitting a packet of 1500 byte requires 120 ns, the *inter-packet time*. This time is proportionally shorter for smaller packets, e.g., 10 ns for a 128 byte Voice over IP VoIP packet. Compared to the clock frequency of current networking switches, which is $\sim$ 1 GHz $\widehat{=}$ 1 ns/clock cycle, this is a very short time [3]. Imagine that sending and receiving small packets may happen at all 64 switch ports in parallel.

The actual latency of networking switches is significantly higher than the inter-packet times, as many packets are processed in parallel by the switch pipeline. However, we assume the port-to-port switch delay is below 1 μs depending on the concrete switch model [17]. During normal operation, i.e., no failure or congestion occurs, the packet loss is typically zero.

As a consequence of this good performance, the benchmarking equipment must be able to operate at these high link speeds and concurrently must provide a very high time and loss measurement accuracy. Specifically, an exact packet loss detection and a time accuracy of a few nanoseconds are mandatory.

R. Kundel, *Accelerating Network Functions Using Reconfigurable Hardware*,
Springer Theses, https://doi.org/10.1007/978-3-031-52872-9_4

As discussed in Sect. 2.5, existing commercial and open-source approaches do not provide this accuracy, especially not in a flexible environment. Note that existing commercial tools build on FPGAs, operating on a lower internal clock frequency than networking switches [19]. This implies a reduced time accuracy as the inter-packet time of small packets is in the range of a single clock cycle.

In this chapter, we discuss how highly accurate and flexible network function testing can be realized with programmable network hardware. This approach allows a sophisticated understanding of the tested hardware, allowing further optimizations of Internet access networks and computer networks.

Requirements on the Testing System

From the above-mentioned demands and challenges, we derive the following requirements for a testing and benchmarking system:

- **Flexible and Easy Load Generation:** Generating complex traffic patterns in hardware is challenging, especially if the sender should react to incoming packets. Therefore, a load generation should be possible in a any way, e.g., utilizing existing standard software tools.
- **Throughput:** It is mandatory to generate test loads up to 100 Gbit/s at any packet size. Current software-based load generators [4] or trace replay tools [1] do not support this. Therefore, hardware assistance is mandatory to achieve high rates. Further, the test load should be precisely adjustable.
- **Time Accuracy:** To capture the behavior of the tested network function in detail, a nanosecond time accuracy for packet timestamping is mandatory. For some analysis, it is compulsory to capture timestamps for every packet, even at 100 Gbit/s.
- **Loss Detection:** In order to detect packet loss, each packet sent in and received from the tested network function must be captured. This loss detection requires exact packet counting, as the benchmarking system should be able to prove zero packet loss behavior of the tested network function.
- **Extendability:** Every network experiment is different, and they often have particular demands. Therefore, the test framework should be modular and easily extendible, including open interfaces and common data formats.

4.1 Overview on Disaggregated Network Function Testing

We propose the `P4STA` framework to address these requirements, which we will discuss in the following. The overall topology is shown in Fig. 4.1 and consists of several interconnected components, which will be discussed in the following:

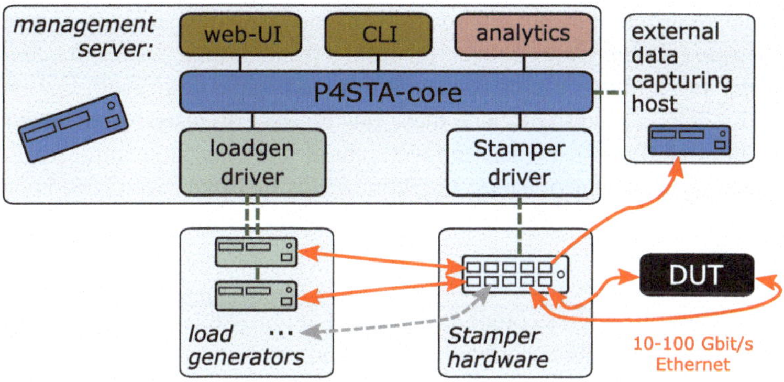

Fig. 4.1 Disaggregated load generation, packet timestamping, and experiment orchestration in the P4STA framework

The **Device Under Test (DUT)** is the network function to be tested. We consider the DUT as a black box, as the internals of the DUT are not necessarily known, and this knowledge is not required.

The **load generators** are commodity servers with Network Interface Cards (NICs). This allows the execution of any software-based packet generator, e.g., *MoonGen* [5], *TRex* [4], *tcpreplay* [1] or *IPerf* [10]. Thus, the complete flexibility of computer software is usable, including congestion-controlled transport protocols and the verification of received packets. The number of load generator servers is variable and allows the realization of complex measurement scenarios, such as generating many parallel subscriber sessions and maintaining the belonging state.

The **Stamper** is the core component of the P4STA measurement setup. It is responsible for establishing the connection between the load generators and the DUT, utilizing commodity Ethernet links. Further, high-precise timestamping, load regulation, and packet loss detection are performed within the Stamper hardware. Later, in Sect. 4.2, this core component of P4STA will be discussed in detail.

The **external data capturing host** (short: *external host*) is an information sink and temporary storage during network experiments. The *Stamper*, unable to store large amounts of data, sends selected network packets with additional metadata information to the *external host*. The *external host* is a software application that receives these packets and extracts the information to be stored, e.g., packet timestamps. After the experiment, this information is provided as an aggregated file, e.g., comma-separated values.

The **management server** is the central control entity in P4STA. All other components of P4STA, i.e., the *load generators*, the *Stamper*, and the *external host*, are connected to it. This server enables a centralized configuration, measurement execution, results collection, and analysis. The *management server* can be executed on any server within the testbed as no special requirements exist. Its main component is the *P4STA-core*, which provides all functionality and generic internal driver APIs for the *Stamper*, *load generators*, and the *external host*. These APIs allow an easy exchange, for example, of different *Stamper* hardware realizations. After a measurement experiment, the *management server* collects the results from the *load generators*, *Stamper*, and *external host* and stores them persistently, including the setup configuration. The *analytics* module is used to evaluate these raw data, as discussed later in Sect. 4.5. A Command Line Interface (CLI) and web interface are used to control the P4STA core, including all its functionality. In addition, the analyzed results can be visualized in the web interface, and all data can be downloaded for further processing in third-party tools.

Note that this disaggregated architecture does not imply that a dedicated server instance is required for every component. Instead, it is possible to run all components in a single P4-programmable hardware switch and its co-located management CPU.

The complete P4STA framework is open-source available, including three different *Stamper* implementations.[1]

In the following Sect. 4.2, we will describe the P4STA *Stamper* in detail. In Sect. 4.3, the postulated accuracy and performance of this approach are evaluated. Next, in Sect. 4.4 and Sect. 4.5, we introduce the general workflow of this framework and fundamental analytics features, enabling fast network function testing. Last, we discuss the presented approach and evaluation results in Sect. 4.6.

4.2 Functionality of the P4STA Stamper

The *Stamper* is the central component of *P4STA*. It is the only component of *P4STA* relying on specific hardware, as it realizes the performance and time-critical functions. In this work, we focus on the following three P4 platforms as a basis for the *Stamper* realization:

- **bmv2:** The behavioral model v2 (*bmv2*) is the software reference implementation of a P4 switch provided by the P4 language consortium [16]. Although this P4 switch does not provide the required latency and bandwidth performance, we used it as the architecture for the *Stamper* reference implementation. Here, no hardware-specific constraints must be considered, and teh *bmv2* can be very well integrated into emulated test networks for development purposes and automated testing.
- **Intel Tofino:** The *P4-Tofino* platform offers up to 64 Ethernet ports, each offering 100 Gbit/s link speed [9]. The internal switch pipeline can be described by the P4 programming language. This platform is the primary hardware target of *P4STA*.
- **Netronome NFP-4000:** This P4-programmable smart network interface card (*P4-SmartNIC*) provides either two 40 Gbit/s ports or two 10 Gbit/s ports. It must be integrated into a server due to its PCIe form factor. Further, it is a less costly solution than a P4-programmable Tofino. Therefore, it is well suited for a low-budget *P4STA* setup in a single server if the lower port count and data rates are sufficient.

[1] https://github.com/ralfkundel/p4sta

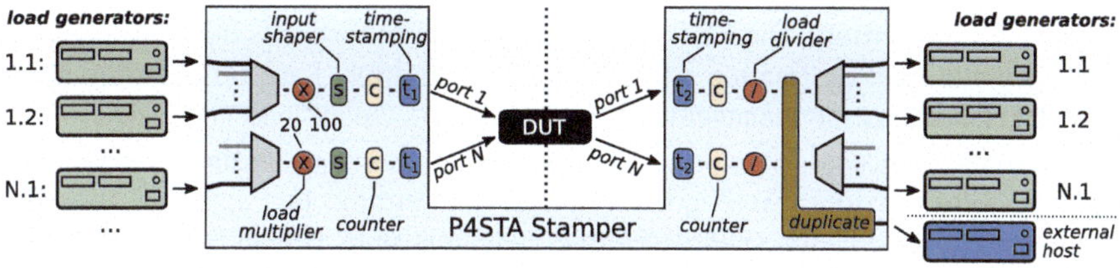

Fig. 4.2 Internal functional components of the Stamper realized with P4-programmable hardware, attached to load generators for packet generation and verification

However, the functional behavior of the three *Stamper* implementations is similar and will be presented in the following sections. Figure 4.2 visualizes the internal components of the *Stamper* device. The packet flow in this representation is only from left to right. Each load generator server is shown twice, (1) as a packet sender on the left side and (2) on the right as a receiver. Analogous, each port of the DUT is represented twice: for ingressing and egressing packets. This means that every packet traverses the *Stamper* device two times: before and after passing through the DUT (compare Fig. 4.1 for the overall scenario).

4.2.1 Load Generation and Aggregation

P4STA supports the usage of multiple software-based load generators in parallel and numerous connected ports of the DUT. This parallelism allows the generation of many different stateful packet streams and the validation of each packet after the DUT. The load generation servers are divided into N groups, consisting of one or multiple load generators each. Each generator group is assigned to one of the N input ports of the DUT. The *Stamper* aggregates all packets sent from one of these load generation servers on this DUT port.

Note that a load generator is characterized by an Ethernet port, i.e., a single physical server with multiple ports can run multiple parallel load generator instances.

After passing through the DUT, the packets enter the *Stamper* device again. Now, multiple packets arriving on a single DUT port must be forwarded to the corresponding load generator server. For example, a packet arriving at port 1 can be forwarded to load generator 1.1 or 1.2.

To cover most scenarios and allow future extensions, we realize three different and concurrent forwarding mechanisms, all implemented in P4STA:

- **Layer 1 forwarding:** This forwarding mechanism determines the egress port only based on the DUT ingress port, e.g., the *Stamper* port 3 will be forwarded to port 7. However, it allows only a single load generator per DUT port. On the other hand, this mechanism is robust against packet corruption, i.e., no valid destination information is provided by the packet.

- **Layer 2 forwarding:** The L2 forwarding mechanism forwards the packets based on their Ethernet destination address to enable multiple load generation servers per DUT port. As the management server has access to all load generators, it can retrieve the dedicated L2 addresses and install the flow rules in the *Stamper*.
- **Layer 3 forwarding:** Analogous to the L2 forwarding, this approach uses the packet destination address to determine the egress port. However, the IPv4 destination address is used instead of the Ethernet destination, allowing multiple receiver addresses on the same port.

A fourth forwarding logic can be simply added if a more advanced forwarding mechanism is required, e.g., a hashing-based packet distribution. Note that this is possible (1) due to the modular design and (2) as the source code of the framework is available.

4.2.2 Load Multiplication, Traffic Shaping and Loss Detection

In addition to the load aggregation mechanism introduced in the previous section, it is beneficial in some scenarios to multiply the generated load in hardware.

Specifically, a software-based load generator creates a packet and sends it to the *Stamper*. Next, the *Stamper* duplicates this packet several times, e.g., 20 or 100 times in the example of Fig. 4.2, and sends all packets towards the DUT. For example, a software load generator can replay a packet trace with 1 Gbit/s; however, the DUT experiences an input rate of 100 Gbit/s. This is a powerful functionality, as a single server is sufficient to generate high test loads.

This load multiplication leads to massive packet duplication, which can disturb the receiving load generator on the right side. For example, the receiver of a stateful TCP load generation would detect duplicated packets, and the congestion control reduces the sending rate as a consequence. To overcome these effects, the P4STA *Stamper* filters all duplicated packets after the DUT, denoted as *load divider* in Fig. 4.2. A flag in the P4STA header, introduced in the following Sect. 4.2.3, indicates whether a packet is the original one or a duplicate. Based on this information, the *Stamper* can easily realize the *load dividing* functionality, i.e., dropping all duplicated packets behind the DUT. In the case of packet loss in the DUT, this applies proportionally to the original packets and is detectable by the congestion control of the load generator.

The capability to test a DUT at a precisely adjusted rate is crucial. However, a precise rate-limiting in software is challenging, especially when load duplication is used. Therefore, the P4STA *Stamper* allows a hardware-based traffic shaping for each egress port towards the DUT. Traffic shaping describes the rate-limited dequeueing of packets from a queue. Later, in Sect. 4.3.2, we will investigate the traffic shaping on the *P4-Tofino* platform and compare it with a software-based rate limiter. Note that traffic shaping in P4STA is currently only supported on the *P4-Tofino*, realized by a token-bucket limiter of the egress queue. As the queues are realized in this architecture between the ingress and egress pipeline, packet timestamping and counting can be performed afterward. In contrast to this, the investigated *P4-SmartNIC* platform offers no programmability after the egress queues. Counting the packets before these queues, which drop packets because of enabled rate limiting, leads to erroneous results. Therefore, no packet loss detection could be performed on this platform if traffic shaping is enabled.

The *Stamper* must count packets before and after the DUT to address the goal of exact packet loss detection. P4-programmable pipelines allow a flexible packet and byte counting of different flows. Within the *Stamper* device, we utilize these capabilities for two different counters:

- **Port counter:** These counters detect every packet received or sent on any port, including the load generator ports and the external host. Therefore, they are very well suited for testbed monitoring and debugging. However, software-based load generators tend to send background noise, e.g., IPv6 neighbor discovery packets dropped by the DUT, making this counter type unsuitable for exact loss detection.
- **P4STA-flow counter:** In contrast to the port counters, this counter captures only packets belonging to flows measured by P4STA. For example, *P4STA* can be configured to capture only *UDP* or *TCP* packets.

4.2.3 Packet Timestamping

Latency measurements and, consequently, packet timestamping is a fundamental task of the P4STA *Stamper*. Every packet must be timestamped before and after the DUT, as shown in Fig. 4.2. As a result, a series of timestamp tuples for each packet, t_1 and t_2, is generated. These timestamps can be used to compute the latency of a single packet, and to compute an inter-packet time of consecutive packets.

Two general questions must be answered regarding the packet timestamping:

(1) Where the timestamps should be stored, and (2) how to retrieve these timestamps.

4.2.3.1 Timestamp Data Storage

Storing two timestamps for each packet at line rate is very challenging. If we assume a rate of 10^8 packets per second and 16-byte measurement data per packet, the *Stamper* creates 1600 MB/s data. However, the internal memory of modern network switches provides only a few MB capacity, as it is realized with fast but small Static Random Access Memory (SRAM) (compare Sect. 2.5). Therefore, only little data can be stored there, e.g., the previously mentioned counter values or the timestamps of packets in flight. Storing all timestamps of an experiment in the *Stamper* hardware is not possible, even if the experiment takes only a few seconds.

In general, two approaches for storing timestamps exist:

- **In the Stamper:** The timestamps are taken by the *Stamper* device and stored internally. The switching hardware does not provide the required capacity, so the data must be transferred to the control CPU. In addition to every timestamp, packet classification information must be stored to assign the two timestamps later. Both transferring data to the control CPU at high rates and packet classification are challenging.
- **Inside the Packets:** P4-programmable hardware is made for packet header modification, including adding additional headers to the packet. This method allows the storage of timestamps and further data within the packet headers. The challenge to match the two timestamps on a single packet does not apply, as every packet carries its own timestamp. Therefore, we chose this approach, and its realization is presented in the following.

Additional header space is required to store the timestamps. By default, network protocols do not allocate any unused header fields by default which can be used.

However, the TCP protocol allows header extensions, which can be added either by the sender or by an intermediate network node, i.e., the P4STA *Stamper*. We utilize the option ID $0x0f$, which TCP does not use. The option has a total length of 16 bytes, 2 bytes to advertise the option ID and length, and the remaining bytes as payload. We name this extension the *P4STA header* in the following, and the header structure is shown in Fig. 4.3.

When a packet passes through the *Stamper* device the first time, the *P4STA header* is added between the existing TCP header and the packet payload. For this, the *Stamper* updates the length and checksum fields of the IP and TCP headers. The TCP checksum is updated incrementally, which means the old checksum is used as the start value, and all packet modifications are applied to it. Further, the *Stamper* can already set the *timestamp_1* value. The second timestamp remains zero. Each timestamp is stored with a accuracy of 1 ns as unsigned integer value. The *flags* are used for additional packet metadata. For example, during load duplication, the

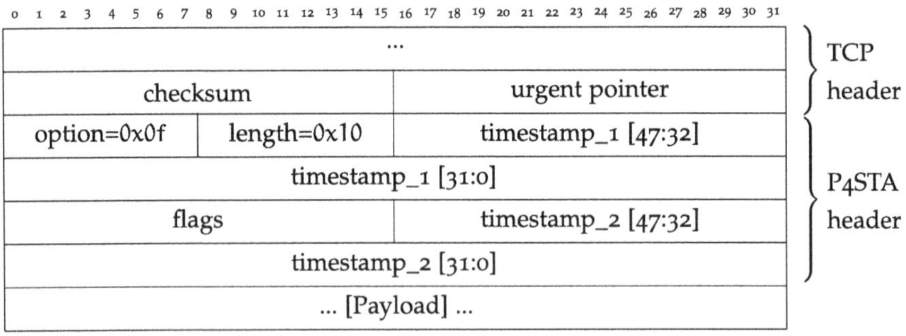

Fig. 4.3 TCP header extension for P4STA, storing the timestamps before and after passing through the DUT with 1 ns precision. Analogous, the timestamps are encoded in the UDP payload. Figure derived from: [12]

packets are marked as duplicates by a flag to identify and filter them later. Further, the flag bits can be used to encode the ingress port of the DUT if more than a single port is used.

When the packet passes through the *Stamper* the second time, the *timestamp_2* is taken and added to the appropriate field in the *P4STA header*.

This approach works only for TCP packets for apparent reasons. Therefore, we extend this approach to store the *P4STA header* in the first 16 bytes of the UDP payload if the packet size is sufficiently large. In contrast to the TCP approach, this does not increase the packet size; however, it corrupts the payload and can only be used if this is no drawback.

Finally, after passing through the DUT, the packet exists within the *Stamper* and the two timestamps are encoded in the *P4STA header*. The information is collected by the *external data capturing host* and is described in the following Sect. 4.2.4.

4.2.3.2 Locality of Timestamping Logic

One reason to utilize P4-programmable hardware for the *Stamper* functionality is the potential of timestamping packets with high accuracy. However, multiple places in the pipeline of the utilized hardware exist, where timestamps can be retrieved.

We can generally formulate the following rule of thumb: Packets should be timestamped as late as possible before the DUT and as early as possible after the DUT.

Figure 4.4 depicts the internal pipeline of the *P4-Tofino*, one of the two hardware targets for the *Stamper*. The Tofino architecture allows two different ways to retrieve timestamps: (1) In the P4-programmable ingress and egress pipeline, marked as *stage 1–stage n*, or in the ingress and egress Media Access Controllers (MACs) of each port. Note that the MACs are not programmable by P4; however, packet timestamping for PTP time synchronization can also be used.

Following the before-stated rule of thumb, we expect a higher accuracy for the MAC timestamping. Nevertheless, we evaluate both approaches as part of the evaluation in the following Sect. 4.3.1.

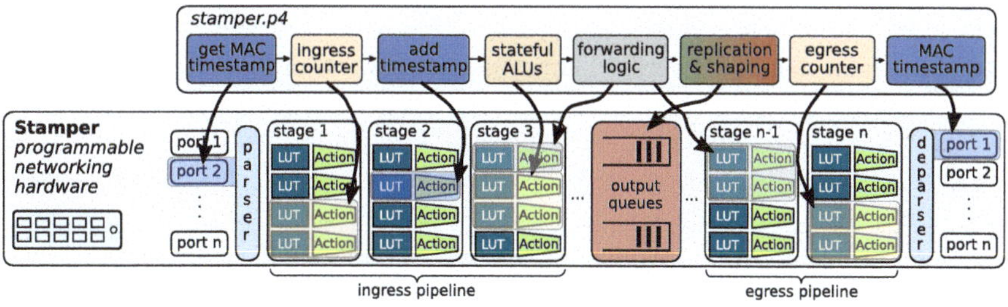

Fig. 4.4 Mapping of the P4STA Stamper functionality, described in *stamper.p4*, on the P4-programmable hardware pipeline of the Intel Tofino. ©2021 IEEE. Reprinted, with permission, from "Network Testing Utilizing Programmable Network Hardware" [13]

Packets sent towards the DUT are timestamped in the egress MAC. The egress MAC stores the timestamp directly into the *P4STA header* and updates the TCP checksum accordingly. For this, the P4 pipeline computes the exact position of the timestamp and the TCP checksum within the packet. Then, the egress MAC writes the current time on this position in the packet and updates the checksum field with the timestamp value.

Timestamping of packets after the DUT works slightly differently. The ingress MAC takes the timestamp and provides it as metadata to the P4 pipeline. Later, in *stage 2* of the depicted pipeline, this timestamp is written into the *P4STA header*.

All other *stamper.p4* implementation functionality is mapped on the P4-programmable pipeline parts.

The *P4-SmartNICs* used in this work offer the same MAC timestamping capabilities for time synchronization and are used by our implementation.

4.2.4 Data Acquisition in the Data Capturing Host

The measurement data must be stored after a packet traverses the *Stamper* the second time and the second timestamp was taken. To accomplish this, the *Stamper* duplicates the packet, sends the original to the receiver, and the copy is forwarded to the *external data capturing host* (compare Fig. 4.2). The *external host* extracts the two timestamps from the *P4STA header* and stores them in a comma-separated values list. In addition, the packet size is stored. Further information, e.g., the packet IP, can be extracted similarly if the information is required. However, every additional extracted and stored information causes a significantly increased memory demand.

The processing within the external host is not time-critical, i.e., the information content remains unchanged whether a packet is parsed immediately or in 1 ms later.

Sending packets at a very high rate to the external host can quickly overload the software process, and uncontrolled packet loss occurs. The *Stamper* can send only every M_{th} packet to the *external host* to counteract this. In contrast to the uncontrolled packet loss, which typically occurs in batches, this behavior is very deterministic. For most evaluations, timestamps for every M_{th} packet are sufficient. Note that this factor does not influence the packet loss detection within the *Stamper*. Further, the maximum, minimum, and average latency is captured within the *Stamper* hardware, not influenced by this.

The *P4STA-Core* starts and stops the *external host* automatically. After an experiment, the results are stored as comma-separated values and copied to the *management server* for the analysis of the results.

The *P4STA* framework provides two implementations of the *external host*, both providing the same functionality:

- **Python External Host:** This implementation opens a raw socket on the network interface of the *external host*. A raw socket receives every incoming packet on the bound interface, including Ethernet, IP, and transport layer headers. By this,

the python implementation has access to the *P4STA header*, can extract the information, and store them in an internal data structure. Raw sockets can be opened on almost any Linux system without modifications, making this approach uncomplicated. However, raw sockets have a comparably bad performance in receiving packets. We observe a maximum receive bandwidth of $\leq$ 1 Gbit/s, depending on the packet size and server CPU. Using another programming language would have a slight performance influence, but receiving every packet at high rates would still not be possible [14].
- **DPDK External Host:** The *Data Plane Development Kit (DPDK)* is a user-space driver framework for network interfaces, aiming for the highest possible performance. The entire incoming packet is handed over to the user space application, identical to the raw socket behavior. This application is written in the programming language C, offering the highest achievable performance. The DPDK driver requires supported network interfaces cards. Most NICs with $\geq$ 10 Gbit/s link speed support this driver framework, and thus this is no major limitation. However, a DPDK-specific kernel module must be compiled and loaded for the used NIC port, making the *DPDK external host* a more error-prone implementation. This module must be rebuilt after every kernel update.

Consequently, the *python external host* offers an easy-to-use solution; however, the DPDK realization should be used if the timestamps of every packet should be stored. In our tests, we experienced zero dropped packets for the DPDK driver up to 10 Gbit/s, even at a small average packet size of 200 bytes. Higher data rates are not yet tested at the external host. The *Stamper* counter also captures all packets sent to the *external host*. These counter values allow the *P4STA-core* to compare the number of packets sent to the external host with the received packets. The derived difference indicates uncontrolled packet loss and is a sign of non-trustworthy results of the *external host*. To that end, we introduce the downscale factor M, indicating how many packets are sent to the *external host*, i.e., every M_{th} packet. This factor M should be increased until zero packet loss between the *Stamper* and *external host* occurs.

4.3 Accuracy and Performance Evaluation

In this section, we evaluate the *P4STA* framework concerning the aimed performance goals. First, in Sect. 4.3.1, we focus on the time accuracy of the *Stamper*. Second, in Sect. 4.3.2, the rate-limiting behavior of the hardware-based traffic shaping in the *Stamper* is evaluated. For the rate-limiting, we focus on packet burst occurrence while traffic shaping.

4.3.1 Time Accuracy

To investigate the accuracy of the time measurement mechanism, we use a DUT with known and constant latency: a fiber-optic cable. While the speed of light in a vacuum is $\sim 300 \cdot 10^6$ meters per second, the propagation speed in optic cables is lower but still constant and known. Consequently, the *Stamper* device, which takes the timestamps, is responsible for any potential variance in the measurement results.

In the following, we investigate the *Stamper* implementation realized on the *P4-Tofino* platform. In this architecture, the timestamping can be realized in two different ways (compare Sect. 4.2.3 for details):

- **P4 timestamping:** The incoming and outgoing packets are timestamped in the P4-programmable ingress or egress pipeline, respectively.
- **MAC timestamping:** The Media Access Controllers (MACs), the closest functional component to the fiber-optic cable, can timestamp packets, i.e., before the ingress parser and after the egress deparser.

The experiment setup is as follows: (1) A load generator creates 1514 byte test packets with a rate of 10 Gbit/s for 10 seconds and sends them into the *Stamper*. (2) The *Stamper* timestamps the packets and forwards them into the DUT, the fiber-optic cable. This link is configured as 40 Gbit/s Ethernet. (3) Next, the *Stamper* receives the packets, and the second timestamp is added. (4) All packets are sent to the receiving load generator, and the *external host* captures every 50_{th} packet.

We repeat this experiment for multiple fiber-optic cables of different lengths. Each run consists of multiple million packets and is therefore significant.

The numeric results of this experiment for both timestamping methods and all tested cable lengths are shown in Table 4.1 and visualized in Fig. 4.5. Note that it is essential to determine the real cable length as the manufacturer information is imprecise, e.g., the 3 m cable in our tests has an actual length of 3, 18 m.

The average latency values for the *P4 timestamping* approach are significantly higher than for the *MAC timestamping*. This deviation corresponds to the known internal architecture of the *P4-Tofino* (compare Fig. 4.4). As the *MAC timestamping* approach retrieves the two timestamps closer to the DUT, the lower latency is self-

Table 4.1 Measured latency and standard deviation for MTP OM4 multi-mode fibers at 40 Gbit/s link speed. Zero packet loss in all measurements

	Actual Length	P4 timestamping		MAC timestamping	
		Avg. latency	Std. dev.	Avg. latency	Std. dev.
1 m	1.06 m	388.80 ns	1.53 ns	107.83 ns	1.46 ns
2 m	2.08 m	393.70 n	1.74 ns	112, 85 ns	1.61 ns
3 m	3.18 m	399.14 ns	1.69 ns	118.34 ns	1.52 ns
10 m	10.12 m	433.46 ns	1.64 ns	152.88 ns	1.61 ns

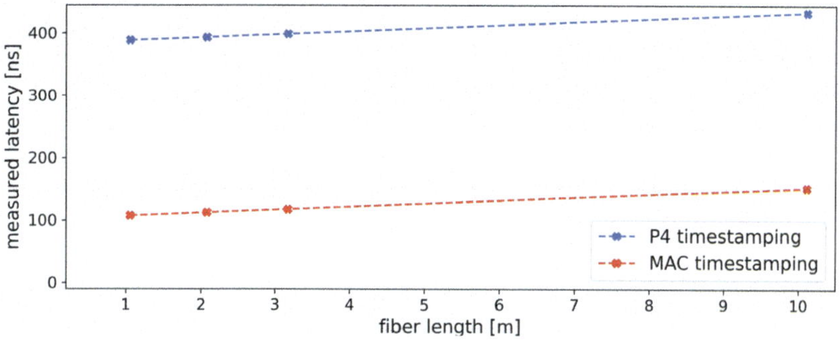

Fig. 4.5 Measured latency of the fiber-optic cable as DUT with P4STA for the *P4-Tofino* Stamper. The two curves represent the two stamping options in the switch pipeline. This plot visualizes the results in Table 4.1

evident. We can state that the *MAC timestamping* is superior to the *P4 timestamping* and should be used.

Further, we observe a similar and low latency standard deviation for both timestamping methods, i.e., always below 2 ns. A significant improvement of the *MAC timestamping* approach cannot be determined; however, this may change if more *Stamper* ports are used in parallel.

The visualization in Fig. 4.5 indicates a linear correlation between the cable length and the measured latency. We perform a linear regression on the measurement values to confirm and substantiate this relationship. A linear equation is represented as follows:

$$f(x) = a \cdot x + b$$

In the particular scenario of this thesis, $f(x)$ describes the measured latency, x the cable length, a the latency per meter cable, and b is a constant offset.

After applying the measurement values to the linear regression mechanism, we get the following results: For *P4 timestamping*, the basis latency is $b = 383.49$ ns, and the propagation speed is $a = 4.94$ ns/m. The average absolute error is 0.05 ns. For *MAC timestamping*, the basis latency is $b = 102.53$ ns and the propagation speed is $a = 4.97$ ns/m. Here, the average absolute error is 0.02 ns.

Both results are excellent; however, the *MAC timestamping* approach stands out positively. The speed of light in a fiber-optic cable depends on the fiber's refractive index. According to related work, a propagation speed between 67% and 69% of the speed of light in vacuum is realistic, i.e., 4.83 ns/m to 4.98 ns/m [18, 20]. Consequently, these numbers match with our measurement results and illustrate the desired accuracy of the P4STA approach realized on the *P4-Tofino* platform.

However, it is required to perform an **offset correction** while performing experiments with real DUTs, especially if the DUT latency is low. For this, the constant latency factor must be determined for every experiment, consisting of the constant latency n and the propagation delay of the cable. Since these values are known, they can be subtracted from the latency values after the experiment. Note that the constant latency offset n depends on the link speed of the DUT-cable and must be determined for every combination of *Stamper* hardware and DUT-port configuration.

4.3.1.1 Influence of the Hardware Platform

Besides the *P4-Tofino* platform, we presented the *P4-SmartNIC* as a secondary *Stamper* implementation. We repeat the previous experiment, measuring the speed of light in cables, with this platform. In Table 4.2, the results for the *P4-SmartNIC* and the *P4-Tofino* are shown. The *P4-Tofino* utilizes a 100 Gbit/s cooper cable of 2-meter length as DUT, configured as 10 Gbit/s or 100 Gbit/s Ethernet connection.

For the *P4-Tofino* platform, we can observe a slight difference between the two link speeds but also regarding the basis latency in contrast to a fiber link (compare Table 4.1 for the base latency of a 40 Gbit/s fiber connection).

The results of the *P4-SmartNIC* are significantly worse. While the higher average latency is a constant factor that can be subtracted, the higher latency standard deviation leads to measurement inaccuracies. Yet, $\sim$ 8.4 ns is still a very low value. Figure 4.6 visualizes the difference between these two approaches. The inaccuracy of the *P4-SmartNIC* compared to the *P4-Tofino* becomes clear by the more widespread vertical distribution.

In these experiments, the *P4-SmartNIC* detected zero packet loss, which means that the packet counting before and after the DUT works appropriately.

We can conclude that the *P4-SmartNIC* is a well-suited alternative to the *P4-Tofino* switch if the two physical ports are sufficient and not the highest possible timestamping accuracy is required. Note that the software load generators must be realized within the system of *P4-SmartNIC*, as the DUT uses the two available ports. Similar cards with 2x 40 Gbit/s ports exist, which can be configured to 8x 10 Gbit/s, allowing an external load generator. However, these cards were not available during this work.

Table 4.2 Latency over time for a (fiber-optic) cable as DUT at 10 Gbit/s throughput and 1500 byte packets for the two Stamper implementations and a varying link speed

	Link Speed	Avg. Latency	Latency Std. dev.	Max. Latency	Packet Loss
P4-Tofino	10 Gbit/s	182.96 ns	1.93 ns	193 ns	0
P4-Tofino	100 Gbit/s	150.56 ns	1.10 ns	165 ns	0
P4-SmartNIC	10 Gbit/s	364.33 ns	8.41 ns	392 ns	0

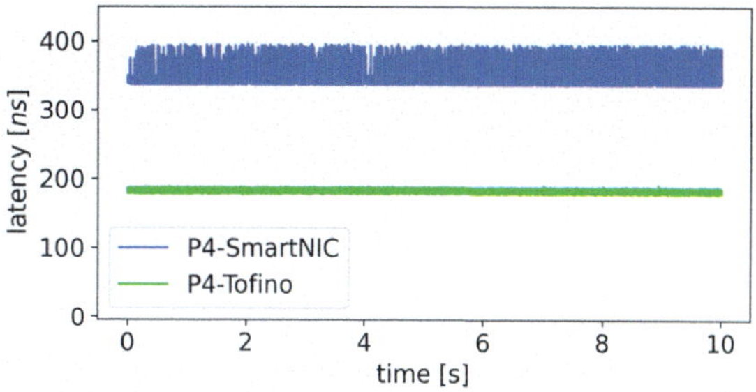

Fig. 4.6 Latency over time for a (fiber-optic) cable as DUT at 10 Gbit/s link speed, 10 Gbit/s throughput, and 1500 byte packets. This representation extends Table 4.2

4.3.2 Microbursts in Traffic Load Generation

Utilizing the *load multiplication* of the *Stamper* device allows reaching a test load of up to 100 Gbit/s even for small packet sizes and a low input rate. Recently released P4-programmable network switches support even up to 400 Gbit/s link speed on each port [9]. We expect no change in the behavior of the proposed concepts for a four times increased link speed; however, we could not investigate this novel hardware within this work.

However, achieving a test load at link speed is not always desired. For example, testing a DUT on specific or different input rates can be very insightful, i.e., an input rate parameter sweep. For this, the *Stamper* allows traffic shaping at a given rate for each egress port towards the DUT.

An ideal traffic shaping mechanism would send only one packet at once, with a constant inter-packet time. If multiple packets are sent at once in small microbursts, the time between two bursts is longer to achieve the configured rate. In general, if a DUT should be tested at a specific input rate, no large microbursts are intended and may negatively influence the results.

In the following, we compare the traffic shaping accuracy of the *P4-Tofino* platform, used as *Stamper* in this work, with the widely-used software load generator *IPerf3* [10]. For this comparison, we configured a rate of 624 Mbit/s in both investigated platforms. For the *P4STA* approach, a software load generator sends a 1 Gbit/s packet stream into the *P4-Tofino*, which means that the traffic shaping queue is always full. For both scenarios, the Ethernet link is configured to 40 Gbit/s, and *P4STA* captures the packet timestamps for every packet. For this experiment, it is required to capture every packet as the inter-packet times are computed afterward.

The results of this experiment are shown in Fig. 4.7. The two graphs depict the maximum and minimum observed data rate for a window of n packets. The window size n is varied from 100 to 8000.

In general, we observed a bursty behavior for both realizations. The *P4-Tofino* sends packets in bunches of 6 packets, while the *IPerf3* load generator sends hundreds of packets at once. For the *P4-Tofino*, this bunch size depends on the configured link rate and varied in our experiments from 4 to 8. This can be explained by the internal

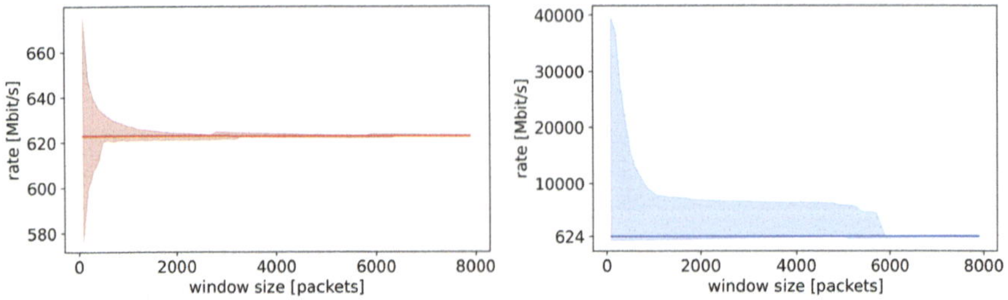

(a) P4-Tofino traffic shaping. (b) IPerf3 traffic shaping.

Fig. 4.7 Average minimum and maximum measured data rate over a window of N packets, ranging from 100 to 8, 000. The rate limit is set to 624 Mbit/s for both realizations. Note the different y-axis scaling

token bucket mechanism, which must enable a rate-limiting of up to the link rate (compare Appendix A.4).

The results in the graph can be interpreted as follows: Considering any series of 100 packets, the maximum observed rate over these 100 packets is $\sim$ 670 Mbit/s, while the lowest rate is $\sim$ 580 Mbit/s. In contrast to this, for the *IPerf3* rate-limiting mechanism, we observed an average rate of $\sim$ 38 Gbit/s for the packet window sizes of 100 and 200 packets. Only starting with a packet window size of $\sim$ 6000 packets, the maximum observed rate is close to the configured rate.

We can conclude that hardware traffic shaping causes significantly lower microbursts than software-based approaches. Note that other load generator implementations may have better performance; however, generating packets in software is easier in batches.

4.4 The P4STA Workflow

Performing network experiments with the *P4STA* framework involves multiple components with different functionality, which must be orchestrated. The *P4STA-core* is the central control entity, managing all other components. In this section, we introduce the automation capabilities of this framework.

During **installation**, all involved servers must be configured to fulfill the needs of the corresponding software components. For example, software dependencies must be installed, and permissions in the operating system must be set. This installation process consists of two phases: (1) All dependencies on the management server are installed. After this, the *P4STA-core* can be started to manage the second installation stage. (2) In the *web-UI*, the server instances for the *Stamper*, *load generators*, and the *external host* must be configured. Following this, the *P4STA-core* installs all dependencies on each server automatically. After this installation process, the *P4STA* framework is ready to use.

The typical workflow of a measurement execution consists of four phases: (1) *Configure*, (2) *Deploy*, (3) *Run*, and (4) *Analyze*.

First, in the **Configure** phase, the entire testbed setup is described. This configuration includes the connectivity of the *Stamper* device, i.e., which ports are connected to the DUT, to the load generators, and to the *external data capturing host*. Further, the IP addresses and credentials for management access of these components are set, and the load generation configuration is described. The configuration can also include *Stamper*-specific configurations, such as supported Ethernet technologies or custom data plane programs extending the default *stamper.p4* implementation.

As an output of the *configure* phase, a JSON configuration file is generated, used in the following phases, and as documentation of the experiment. As distributed systems with many configuration prospects are error-prone, we added a status check for the configuration phase. This check validates all requirements and configured IP addresses on each server, matching with the configuration.

Second, in the **Deploy** phase, the testbed configuration is loaded on the *Stamper* device, and the status of this device and its connections is checked, i.e., if all required Ethernet links are up. This deployment includes the initialization of the required drivers for the selected *Stamper* implementation.

Third, the actual experiment is started in the **Run** phase. The *external host* is started, the *Stamper* counters are reset to zero, and the load generators are started. After the completion of the load generation, the *external host* is terminated. Further, the results from the *external host* and *Stamper* are copied to the central management server.

Last, in the **Analyze** phase, the captured results are evaluated. This includes basic statistics and generous plots of the time series data. By that, the testing engineer can review the first feedback of the measurement within seconds. In addition to the visualization, all raw data, evaluation scripts, and generated results can be downloaded as an archive file. Based on this Python script collection, the user can adapt the visualizations and add further representations of the results. We discuss the details of the analytic functions in the following Sect. 4.5.

All four phases of measurement execution can be controlled either via the *web-UI* or the *CLI*. While the *web-UI* allows an intuitive usage and immediate results visualization, the *CLI* can be used for automated experiment execution. For example, the *MACI* framework can perform sweeping measurement series without human interaction [7].

The *P4STA* framework is a large software framework consisting of many modules, programming languages, and configuration options. Therefore, software testing is essential. We provide multiple test cases covering all phases of the installation and measurement execution for multiple *Stamper* and *external host* implementations. By this, all typical use cases are tested. These tests can be executed in a containerized environment, either locally or in a continuous integration pipeline. However, the testing of the *P4STA* framework is not the focus of this work.

4.5 P4STA Analytics

One essential functionality of *P4STA* is the automated evaluation of results. The *Stamper* device provides counter values for every port and aggregated latency statistics, i.e., maximum, minimum, and average latency. In addition to this, the *external data capturing host* provides time-series data of the two packet timestamps and the packet size. The *analytics* module of *P4STA* evaluates this data and prepares the results in a human-readable way.

The visualization of the *Stamper* results does not require significant processing of the measurement data. In Fig. 4.8, a cutout of the visualization in the *web-UI* is shown for any measurement. The table on the left side presents the number of packets and bytes counted for each ingressing and egressing port of the *Stamper*. Here, the user can choose between all packets or the timestamped packets. On the right side, an auto-generated graphic visualizes the connectivity of the *Stamper* based on the *P4STA* configuration. Further, the packet loss of the DUT in each direction is computed and presented (not shown in the screenshot). Parallel to this representation, all results are exported in human-readable text files. This representation of results assists the user in assessing the quality of a performed measurement much faster than a manual results extraction.

In addition to these basic evaluations, the time series data of the *external host* offers further evaluations, as discussed in the following sections.

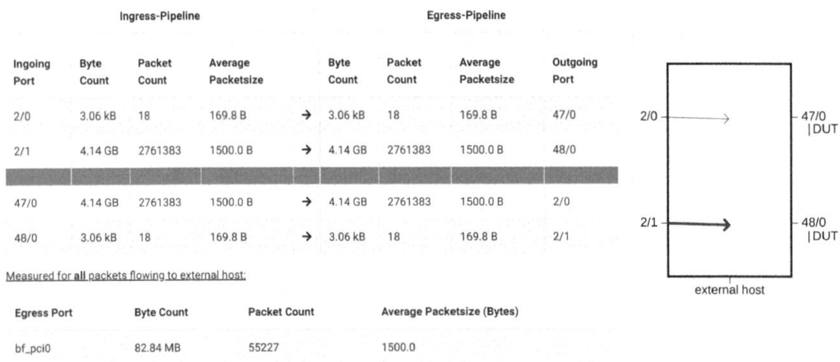

Fig. 4.8 Screenshot of the P4STA web-UI, showing the counter values of the Stamper device as part of the analysis functionality

4.5.1 Latency Measurements

The P4STA framework creates two timestamp series for the experiment, describing the time of a packet before and after the DUT. Based on this data, the latency of each packet can be computed and visualized.

In Fig. 4.9a, the latency of the first 0.9 s after start for four network experiments are shown, performed identically for two different DUTs at two different input rates. The packet size of 1020 bytes is intentionally chosen, as some effects become visible by this. Both DUTs, the DPDK packet forwarder and the FPGA, connect two Ethernet ports with each other and forward every packet. However, these simple network functions are not the focus of this section.

We observe different behavior in the latency curves of the four experiments. The *FPGA*, operating at 9.9 Gbit/s, has a constant low latency. When increasing the input rate to 10 Gbit/s, the same DUT has little packet loss and an increased latency with a saw-tooth pattern. Similarly, the *DPDK* packet forwarder at 9.9 Gbit/s input rate has a constant latency with a visually detectable jitter. The same DUT at an input rate of 10 Gbit/s builds up a queue as the packets cannot be processed as fast as they arrive. However, no packet loss occurs as the buffering capacity of this DUT is larger than the built-up queue in the first 1 s. The numeric results for these four runs are shown in Table 4.3.

In Fig. 4.9b, the latency distribution is visualized. Focusing on the *DPDK* DUT at 9.9 Gbit/s input rate, we can read that $\sim$ 99.8 % of the packets have a latency below 10^1 µs. Only a few packets, the outliers, have higher latency. In contrast to this, the curve for the *FPGA* at the same input rate is straight, which means that no latency outliers exist.

If we compare these observations with the numeric Table 4.3 generated by the P4STA *analytics* function, we can confirm these findings. The latency standard deviation for the *FPGA* at 9.9 Gbit/s is the lowest, an indicator of low jitter. Further, the latency range, i.e., the difference between the minimum and maximum latency,

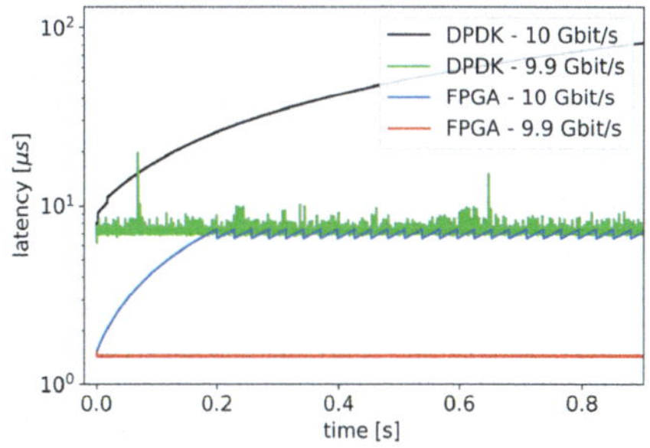

(a) Latency time series.

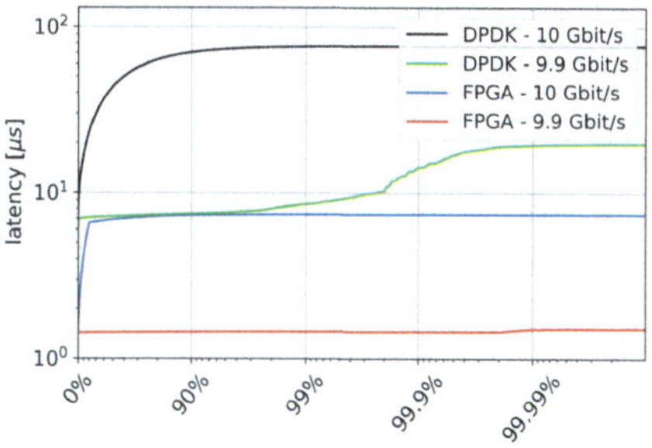

(b) Latency occurrence percentile.

Fig. 4.9 Exemplary measurement results created with P4STA on the *P4-Tofino* Stamper for different DUTs and input rate. The external host captured every 100. packet. Packet size: 1020 byte. No offset-correction

Table 4.3 Numeric measurement results in extension to Fig. 4.9 for the complete experiment with a total duration of ~ 5 s

	FPGA	FPGA	DPDK	DPDK
Input rate:	9.9 Gbit/s	10 Gbit/s	9.9 Gbit/s	10 Gbit/s
Transmitted packets	11887798	12005428	11886608	12007588
Lost packets	0	340	0	0
Loss rate	0%	0.003%	0%	0%
Average delay	1.49 μs	6.89 μs	7.23 μs	338.89 μs
min. latency	1.43 μs	1.45 μs	6.10 μs	6.05 μs
max. latency	1.57 μs	7.43 μs	23.36 μs	733.22 μs
std. latency deviation	28.9 ns	664 ns	427.9 ns	220, 630 ns

is comparably low. These numbers are captured in the hardware of the *Stamper*, considering every packet, not only the random sample captured by the *external host*. In contrast, the *DPDK* DUT investigated at 9.9 Gbit/s input rate has a much higher latency standard deviation and maximum latency value. This underlies the interpretations of the generated graphs.

We can conclude that the multiple numeric and graphical result presentations of *P4STA* can help to interpret the latency behavior of the DUT quickly.

4.5.2 Packet Ordering

Another important field of network function measurement is packet reordering. To investigate this phenomenon, *P4STA* must timestamp and capture every packet. The resulting two timestamp series can be used to detect packet reordering. We expect a strictly monotonous increase of both timestamp series for a network function with no packet reordering.

We analyzed the measurement results of multiple hardware and software devices during our experiments regarding packet reordering. Figure 4.10 presents the results for a DUT that suffers from packet reordering under some circumstances. On the left side, the latency during the 10 s of the experiment is depicted. Red crosses mark a reordered packet. The arrival and departure times of 7 consecutive packets at the DUT are shown on the right side. The red *packet #n* is the marked reordered packet on the left side. One can observe that the seven packets arrive at the DUT with approximately constant inter-packet times. However, the departure times of these packets are not ordered. *Packet #n+1* leaves the DUT before *packet #n*. This reordering can be identified precisely by the automated results analysis of P4STA.

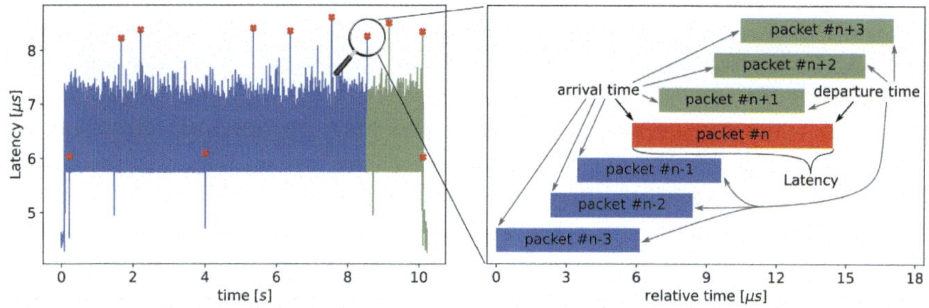

Fig. 4.10 Left: Latency of a DUT over time; red crosses mark reordered packets. Right: Arrival and departure time of 7 consecutive packets at the DUT. Each bar represents the duration a packet stays in the DUT. *Packet #n* is reordered

4.5.3 Network Calculus

In Sect. 2.4, we introduced *Network Calculus* as a framework for describing and understanding the behavior of network functions [6, 15]. The generated time series of timestamp and packet size, created by P4STA, can be used to generate a service curve, following the mathematical instruments of *network calculus*. The depicted service curves in Fig. 4.11 for three different DUTs show the amount of data in byte after a time interval for the first 300 μs and 25 ms after start. At time 0, the first packet of a constant 10 Gbit/s packet flow enters the DUT.

In the case of the *fiber-optic cable* as DUT, we observe a constant bitrate service, starting almost immediately with the first packet. As discussed earlier, the latency of a short fiber cable is rather low, i.e., in the range of ∼ 110 ns including measurement overhead. In contrast to this, the *DPDK packet forwarding* has a constant bitrate service but with some delay. This correlates with our earlier expectations that a DPDK forwarder has a delay of ∼ 10 μs, depending on the packet size. Note that the *fiber-optic cable* in the left and right plot is the same measurement but at different x-axis and y-axis scaling, and serves for comparison.

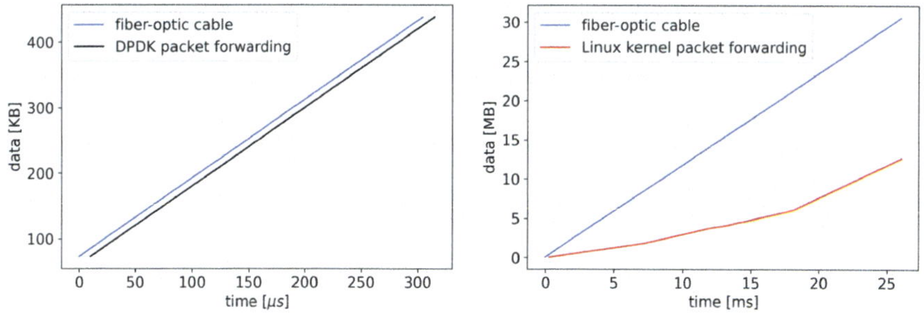

Fig. 4.11 Service curve for different DUTs, on which an abruptly applied load of 10 Gbit/s acts. The packet size is 1500 byte. Note the different x-axis and y-axis scaling

Last, we investigate the behavior of the *Linux kernel packet forwarding*. In this experiment, we observed packet loss that can also be supposed by the gradient of the plot. At the beginning of a busy period, we observe a warm-up behavior for the first 18 ms after which the service rate increases. This warm up behavior may be caused by the software implementation of the Linux kernel and by this CPU caching effects can occur, leading to worse performance in the beginning.

To summarize, we have shown that the mathematical concepts of *Network Calculus* can be applied on the measurement data created by P4STA. An additional value in analyzing the behavior of network functions is created, facilitating the assessment of measurement data. Note that many other analyses can be performed similarly; however, the focus of this work is on the precise measurement data creation only.

4.6 Summary and Discussion

Due to the absence of flexible and performance-oriented test solutions with high time accuracy, we investigated the capabilities of programmable hardware for this purpose in this chapter. The introduced *P4STA* framework combines the benefits of programmable networking hardware and commodity software. By this, high measurement accuracy is achieved while any software-based load generator is used. Simultaneously, high data rates at a low average packet size can be achieved, e.g., 100 Gbit/s at 100-byte packets.

Our evaluation results of the timestamping accuracy have shown that it is sufficiently good to determine the propagation speed of light in fiber-optic cables. The presented algorithms for automated results evaluation allow a timely assessment of measurement results. In the following of this work, we will use the *P4STA* framework to evaluate network functions at the Internet access edge regarding their Quality of Service (QoS) characteristics.

During this work, a similar approach was presented in related work, named "Inband Network Telemetry (INT)" [21], which we already discussed in Section 2.5. Nevertheless, we would like to emphasize at this point of the thesis one major similarity with our work: The proposed INT approach stores the packet timestamps within the network packets, utilizing the timestamping capabilities of P4-programmable network switches, yet, for the use case of network monitoring.

The contributions of this work are not isolated; instead, the *P4STA* framework should be seen as a measurement concept. The generated results provide an ideal basis for further evaluations by data inspection tools, e.g., the traffic analyzing tool *Tranalyzer* [2].

Last, we would like to mention that the company *Keysight Technologies* built a commercial product upon the concepts and results of this chapter [11]. Utilizing the same programmable hardware as this work, i.e., the *P4-Tofino*, they provide a closed-source testing environment. Even though we expect the same accuracy and performance, the advantage of an extensible open-source solution disappears.

References

1. Turner A, Klassen F (2022) Tcpreplay - Pcap editing and replaying utilities. https://tcpreplay.appneta.com/wiki/tcpreplay-man.html. [man page; accessed 18-April-2022]
2. Burschka S, Dupasquier B (2016) Tranalyzer: Versatile high performance network traffic analyser. In: 2016 IEEE symposium series on computational intelligence (SSCI), pp 1–8
3. Chole S, Fingerhut A, Ma S, Sivaraman A, Vargaftik S, Berger A, Mendelson G, Alizadeh M, Chuang ST, Keslassy I, Orda A, Edsall T (2017) DRMT: disaggregated programmable switching. In: SIGCOMM '17, pp 1–14. ACM, New York, NY, USA
4. Cisco Systems, Inc. (2022) TRex - Realistic traffic generator. https://trex-tgn.cisco.com/. [Online; accessed 10-February-2022]
5. Emmerich P, Gallenmuller S, Raumer D, Wohlfart F, Carle G (2015) MoonGen: a scriptable high-speed packet generator. In: Proceedings of the internet measurement conference. IMC '15, pp 275–287. ACM, New York, NY, USA
6. Fidler M, Rizk A (2014) A guide to the stochastic network calculus. In: IEEE Commun Surv Tutor 17(1):92–105
7. Frömmgen A, Stohr D, Koldehofe B, Rizk A (2018) Don't repeat yourself: seamless execution and analysis of extensive network experiments. In: Proceedings of the 14th international conference on emerging networking experiments and technologies. CoNEXT '18, , pp 20–26. ACM, 2018
8. Ghorbani S, Yang Z, Godfrey PB, Ganjali Y, Firoozshahian A (2017) DRILL: micro load balancing for low-latency data center networks. In: Proceedings of the conference of the ACM special interest group on data communication. SIGCOMM '17, pp 225–238
9. Intel, Inc (2022) Intel Tofino 2. https://www.intel.de/content/www/de/de/products/network-io/programmable-ethernet-switch/tofino-2-series.html. [Online; accessed 06-February-2022]
10. iperf3 (2022) https://www.mankier.com/1/iperf3 [man page; accessed 18-April-2022]
11. Keysight Technologies, Inc (2020) Keysight unveils ultra-high-density, multi-terabit test solution for data center operators and network equipment manufacturers. https://www.keysight.com/de/de/about/newsroom/news-releases/2020/ keysight-unveils-ultra-high-density– multi-terabit-test-solution. html. Press Release. Mar. 2020
12. Kundel R, Siegmund F, Blendin J, Rizk A, Koldehofe B (2020) P4STA: high performance packet timestamping with programmable packet processors. In: Proceedings of the IEEE/IFIP network operations and management symposium (NOMS), pp 1–9. IEEE
13. Kundel R, Siegmund F, Hark R, Rizk A, Koldehofe B (2022) Network testing utilizing programmable network hardware. IEEE Communications Magazine 12–17
14. Kundel R, Stiegele P, Tran D, Zobel J, Abboud O, Hark R, Steinmetz R (2021) User space packet schedulers: towards rapid prototyping of queue-management algorithms. In: Proceedings of the international conference on networked systems (NetSys), pp 1–4. EASST
15. Le Boudec JY, Thiran P (2001) Network calculus: a theory of deterministic queuing systems for the internet. Springer
16. P4 Language Consortium (2022) P4 behavioral model (bmv2). https://github.com/p4lang/behavioral-model. [Online; accessed 14-April-2022]
17. Plexxi, Inc (2016) Latency in ethernet switches. White Paper. 2016
18. Singla A, Chandrasekaran B, Godfrey PB, Maggs B (2014) The internet at the speed of light. In: Proceedings of the 13th ACM workshop on hot topics in networks. HotNets-XIII, pp 1–7. ACM, New York, NY, USA
19. Spirent Communications, Inc (2012) Spirent Testcenter Virtual. Product brief
20. Blumenkranz S (2022) What is precisely the speed of light in fiber optics? https://www.quora.com/What-is-precisely-the-speed-of-light-in-fiber-optics. [Online; accessed 16-April-2022]
21. Tan Lizhuang, Wei Su, Zhang Wei, Lv Jianhui, Zhang Zhenyi, Miao Jingying, Liu Xiaoxi, Li Na (2021) In-band network telemetry: a survey. Comput Netw 186:1–24

Chapter 5
Evaluation

Based on our conceptional contributions on *host bypassing*, Internet service creation on programmable hardware, and active queue management in programmable hardware, we will show the viability of these approaches in the following. Further, we will investigate the properties of the prior introduced approaches in detail.

The evaluation of this work builds upon the *P4STA* framework introduced in this work, which allows very accurate measurement of network functions performance. The accuracy of this novel measurement methodology is discussed in the foregoing Chap. 4. First, in Sect. 5.1, we discuss the *host bypassing* approach while presenting measurement results for various deployment scenarios. Following, the concepts for Internet service creation in access networks are discussed in Sect. 5.2, investigating residential and mobile Internet access technologies. To improve the Quality of Service (QoS) in Internet access networks, we proposed concepts for Active Queue Management (AQM) in programmable hardware, which are evaluated and discussed in Sect. 5.3. Last, in Sect. 5.4, we summarize the evaluation results of the contributions and discuss them with the aim of a fast, flexible, and energy-efficient future Internet, relying on programmable hardware.

5.1 Host Bypassing

In Sect. 3.1, we introduced the concept of *host bypassing* for an improved packet I/O in PCIe-based hardware accelerators, i.e., Field Programmable Gate Arrays (FPGAs) and Graphics Processing Units (GPUs). The expected benefit of this approach compared to state-of-the-art is an improved throughput, lower latency, zero packet loss, and less jitter while network packets are transferred between the hardware accelerator and Network Interface Card (NIC). The evaluation results presented in the following two sections will discuss these performance goals while investigating the presented prototypes for *host bypassing* with FPGAs and GPUs, respectively.

R. Kundel, *Accelerating Network Functions Using Reconfigurable Hardware*,
Springer Theses, https://doi.org/10.1007/978-3-031-52872-9_5

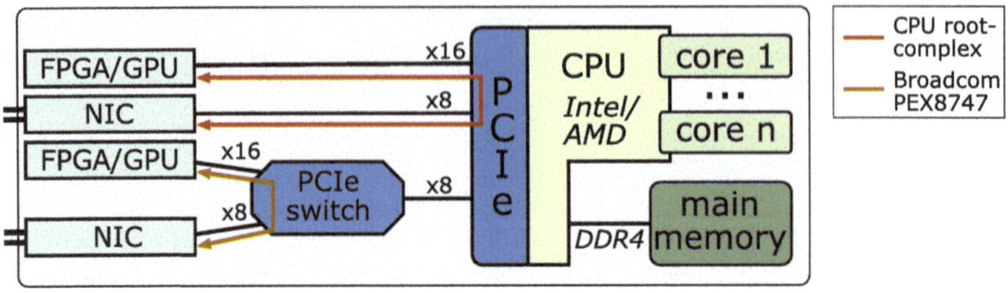

Fig. 5.1 PCIe bus topology utilized while evaluation. © 2021 IEEE. Reprinted, with permission, from "Host Bypassing: Direct Data Piping from the Network to the Hardware Accelerator" [9]

5.1.1 Investigated PCIe Topology

The *host bypassing* approach builds upon PCIe peer-to-peer transfers from one endpoint, i.e., the NIC, to another endpoint, i.e., the GPU or FPGA. For this, it is very likely that the PCIe infrastructure can strongly influence the overall system's performance. Therefore, we investigate multiple PCIe architectures and technologies to identify potential differences between them.

Conventional server CPUs typically provide a sufficient number of PCIe lanes. All peripheral components are connected directly to the PCIe root complex, co-located with the CPU at the same chip. However, the number of PCIe lanes and by that the number of simultaneously connected devices can be increased by a PCIe switch, as shown in Fig. 5.1. The following two CPU models and the PCIe switch will be subject of investigation:

- **Intel Xeon Silver 4110:** This CPU supports up to PCIe Gen 3 support, and main memory is accessible via a DDR4 interface.
- **AMD Epyc 7402:** Similarly, this CPU offers DDR4 interfaces for main memory integration. However, PCIe up to Gen 4 is supported.
- **Broadcom PEX 8747:** This chip is a PCIe switch, supporting up to the third generation (Gen 3) of PCIe. It can be attached to a root complex, and main memory access from an attached endpoint must be forwarded to the PCIe root complex within the CPU. Though, a DMA access on the physical address range of another endpoint can be forwarded only through the PCIe switch, e.g., from the NIC to the hardware accelerator.

5.1.2 FPGA-Based Host Bypassing

In this section, we evaluate the *host bypassing* approach with FPGAs. All evaluation results are generated with the *Xilinx Alveo U50* FPGA accelerator, operating at PCIe Gen 3 and up to 16 lanes (*x16*). Further, this FPGA provides a single 100 Gbit/s Ethernet port, which can be leveraged for comparative evaluation measurements

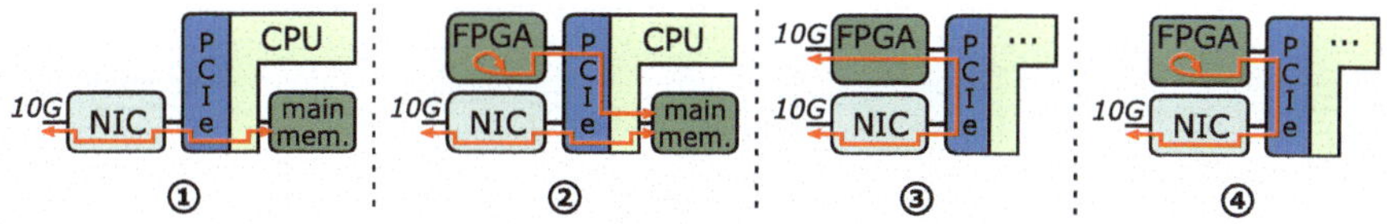

Fig. 5.2 Evaluation scenarios for host bypassing with FPGAs considered in this work. © 2021 IEEE. Reprinted, with permission, from "Host Bypassing: Direct Data Piping from the Network to the Hardware Accelerator" [9]

as described later. Note that this FPGA could be integrated directly into computer networks by this Ethernet port, and *host bypassing* is not mandatory. However, other FPGA accelerator cards do not provide such network connectivity.

In this work, we investigate the following four evaluation scenarios, as shown in Fig. 5.2:

(1) Packets arrive at the 10 Gbit/s Ethernet port of the NIC and are written into the system's main memory via DMA. From there, the packet goes the same path back without any processing and is sent by the NIC. For this, we built upon the sample application `DPDK skeleton`, which is part of the driver framework. The packets are not copied to and from an FPGA accelerator, and this scenario corresponds to packet I/O for classical Network Functions Virtualization (NFV) without hardware acceleration.

(2) The packet data path in state-of-the-art NFV approaches with FPGA accelerators is investigated in the second scenario. In addition to the first scenario, we extend the DPDK application by an FPGA accelerator integration. This extension enables copying the network packets from the main memory into the FPGA and back. First, the receive software process, running on the CPU, hands over memory pointers to the FPGA, indicating the physical address of a received packet. Next, the FPGA can copy this packet via DMA from the system's main memory to the FPGA. Analogous, the software process on the CPU for transmitting packets provides a physical memory address to the FPGA, where the packets to be sent can be stored. After the FPGA confirms the writeback into the main memory, the software process inserts this packet address into the tx ring of the NIC, and the packet will be sent on the NIC's Ethernet port.

(3) The third evaluation scenario describes unidirectional *host bypassing*. This means packets are received via the NIC directly into the FPGA and sent out via the built-in Ethernet port; similarly, the opposite direction receives packets at the FPGA and sends them out via *host bypassing* at the NIC's Ethernet port. This scenario allows an investigation of the rx and tx process independently.

(4) Scenario 4 describes the actual *host bypassing* approach. Packets are received and sent by the NIC, logically attached to the FPGA without copying the data into the system's main memory. Within the FPGA, the packets are not processed by any network function; only the complete I/O procedure is applied to the packets.

Scenarios 2, 3, and 4 are performed with 64 descriptor ring entries for receiving and sending packets within the FPGA, as this parameter shows no influence on the

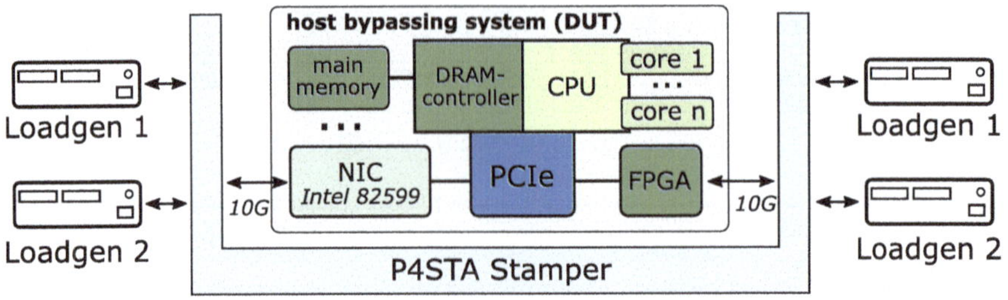

Fig. 5.3 Measurement setup for the host bypassing approach with FPGAs building upon the P4STA framework. "Host Bypassing: Direct Data Piping from the Network to the Hardware Accelerator" [9]

measured performance. For the *DPDK baseline* (Scenario 1), we configured the ring size to 256 entries each, which provides the best achievable performance.

To achieve high test loads and good measurement accuracy, we built upon the *P4STA* framework, introduced in Chap. 4. The overall setup is shown in Fig. 5.3 and consists of two off-the-shelf servers for load generation and one *P4STA Stamper*, more precisely a P4-programmable Tofino switch. Test packets are generated by the load generation servers and sent to the *Stamper*, responsible for load multiplication and traffic shaping to a given bandwidth. Further, the *Stamper* has a direct connection to the Device Under Test (DUT), the host bypassing system to be investigated, and timestamps every packet before and after the DUT to ensure the highest possible measurement accuracy. Packets egressing from the DUT are demultiplexed to the corresponding receiving load generator, which performs a checksum validation. Depending on the evaluation scenario, packets can ingress and egress the DUT via the NIC or the FPGA. All presented latency measurements are corrected by a constant offset for the P4STA setup and cable delays.

5.1.2.1 General Packet I/O Performance

First, we evaluate and discuss the general performance of *host bypassing* with FPGAs, specifically the QoS metrics latency, jitter, throughput, and packet loss. For this, we investigate the host bypassing system (Scenario 3 and 4), and the DPDK baseline (Scenario 1), at a rate of 9.99 Gbit/s with 300−byte UDP packets. This small packet size was intentionally chosen as it causes a higher packet rate (∼4.17 packets per second), allowing us to observe effects in the limiting performance area.

The results in Fig. 5.4 and Table 5.1 show the measured latency in the investigated scenarios. The *host bypassing* measurements (Scenario 3 and 4) are performed with the NIC and FPGA being connected to a PCIe root complex. The NIC and FPGA in the *DPDK baseline* (Scenario 1) and the state-of-the-art approach (Scenario 2), considering the packet I/O into the system's main memory only, are attached directly to the CPU root complex to avoid any performance reduction.

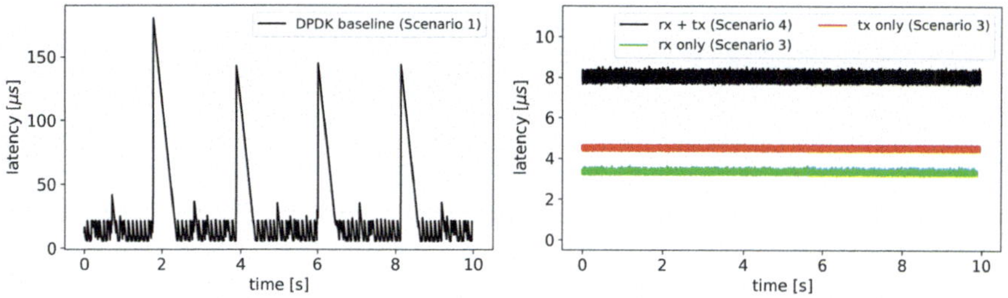

Fig. 5.4 Measured latency for the Scenarios 1, 3, and 4. The test load is 300 bytes UDP packets with a constant rate of 9.99 Gbit/s. Disabled tail pointer update delay, and no tx descriptor writeback. Note the y-axis scaling

Table 5.1 Numeric measured latency of host bypassing, extending Fig. 5.4

	Num. packets	Avg. latency (µs)	Latency std. dev.	Max. latency	Packet loss
DPDK baseline (1)	39,162,180	28.91	35.61 µs	180.26 µs	0
State-of-the-art (2)	39,055,481	168	23.11 µs	1.89 ms	76.31 %
rx host bypassing (3)	36,281,223	3.37	40.01 *ns*	3.37 µs	0
tx host bypassing (3)	38,894,412	4.56	46.96 ns	4.74 µs	0
rx + tx host bypassing (4)	40,124,675	7.99	35.61 ns	8.65 µs	0

The *DPDK baseline* (Scenario 1) receives packets from the NIC into the system's main memory and sends them out without any processing. In this measurement, we already observe high and periodic latency peaks up to 180 ms and an average latency of ~29 ms. Even when ignoring the latency peaks, one can visually observe a high latency jitter. However, this system is able to I/O all packets without any loss or corruption. This scenario is very important as it describes the half path of state-of-the-art packet I/O into hardware accelerators: receiving and sending packets into the system's main memory. Even if the I/O from the main memory into the accelerator causes zero overhead, this behavior is unavoidable. The behavior of the main memory, especially the latency spikes, might be caused by the behavior of the system's main memory controller and the nature of DRAM-based memories, e.g., mandatory refresh cycles, arbitration methods, and shared access with software processes.

Next, we investigate the *state-of-the-art* packet I/O (Scenario 2). Note that this implementation was built by ourselves for comparison purposes and might offer additional optimizations. However, the performance of an ideal implementation would be between our implementation (Scenario 2) and the DPDK baseline (Scenario 1). For the state-of-the-art approach, copying all packets via the system's main memory into the hardware accelerator, we observed massive packet loss, i.e., around 76%. Further, the average measured latency is 168 µs, which is a consequence of the rx buffer in the NIC damming up as the system cannot process the packets faster. We observed less packet loss at larger packet sizes, and we can determine the packet rate as the limiting factor. As the latency is very high due to the bloated rx buffer, we present only the characteristic numbers but no latency over time plot.

In contrast to this, the *host bypassing* approach (Scenario 4) shows up a much better performance, as shown in the right plot of Fig. 5.4. The average latency is $\sim 8\,\mu s$, and only little jitter occurs. Comparing this behavior with the DPDK baseline, an upper bound of the maximum achievable performance for state-of-the-art approaches, the *host bypassing* approach is superior to this. In contrast to the state-of-the-art measurement (Scenario 2), this approach allows a much higher throughput at lower latency and jitter. Therefore, this evaluation shows that the goals of *host bypassing* are achieved.

The detailed behavior of *host bypassing* can be investigated further with the evaluation Scenario 3. In this scenario, packets are received from the NIC via PCIe and sent out at the native Ethernet port of the FPGA (*rx only*) or the other way round (*tx only*). Both directions have lower latency and jitter than the aforegoing bidirectional measurement, which includes both directions. The *rx only* scenario has an average latency of $3.37\,\mu s$, and for the *tx only* scenario, $4.56\,\mu s$. The sum of these two delays is $7.93\,\mu s$, which is 60 ns lower than the bidirectional scenario. This can be explained by the lower PCIe bus utilization and only a very low overhead of the native Ethernet port within the FPGA. For example, if a DMA access of the NIC on the FPGA is ongoing to read a packet to be sent, a parallel DMA read on the rx descriptor ring may be delayed. Further, we can conclude from the results that the *rx* process is causing a higher latency, yielded by the mandatory rx descriptor writeback, notifying the new packet to the FPGA.

5.1.2.2 Non-overloaded Forwarding Behavior

Next, we will investigate the system's performance with a lower packet rate, not overloading the state-of-the-art implementation (Scenario 2). Increasing the average packet size would cause a different latency behavior due to the store and forward nature within the FPGA. Therefore, we lower the sending rate and keep the packet size constant at 300 bytes.

Figure 5.5 depicts the latency over time and latency distribution at 2 Gbit/s for the measurement Scenarios 1, 2, and 4. In all three scenarios, zero packet loss occurs. As expected, the visual examination of the latency over time shows that the *state-of-the-art* approach has a latency characteristic that extends the *DPDK baseline*. Indeed, we still observe latency peaks in the measurements for the *DPDK baseline* and the *state-of-the-art* approach, probably caused by the system's main memory.

From this experiment, we can conclude that also for low rates, the *host bypassing* approach enables significantly lowered latency and jitter for packet I/O. Further, our logically derived assumption that the *DPDK baseline* is an upper bound for state-of-the-art approaches has been fortified by this experiment.

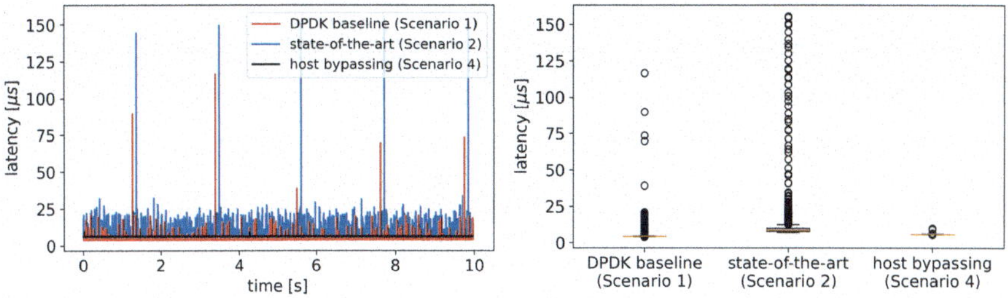

Fig. 5.5 Measured latency for the Scenarios 1, 2, and 4. The test load is 300 bytes UDP packets with a constant rate of 2.00 Gbit/s

5.1.2.3 Influence of the PCIe Topology

In this section, we will investigate the influence of different PCIe topologies on the *host bypassing* approach with FPGAs. Table 5.2 shows the achievable TCP throughput of three concurrent TCP flows, generated with the load generator *iperf3*. The transport protocol TCP adapts its sending rate automatically based on packet loss or corruption. Therefore, the rx queue of the NIC cannot be overloaded, and the average latency stays moderate, even if packet loss occurs. Note that the provided average goodput is the TCP throughput, excluding overhead of L1-L4 packet header fields. A TCP throughput of ~9.29 Gbit/s is the theoretical achievable maximum on a 10 Gbit/s Ethernet link. The test packets are a mixture of large TCP data packets (~1514 bytes) and small acknowledgment packets.

We can observe that for the root complex of the *AMD Epyc 7402* CPU and the *Broadcom PEX 8747* PCIe switch zero packet loss occurs. The end-to-end latency is around 770 ns lower for the PCIe switch, which is a very similar result. The maximum achievable throughput for the *Intel Xeon 4110* CPU is only 7.77 Gbit/s, and packet loss occurs. The average latency is significantly higher than the other two PCIe topologies, as the system is in an overloaded state. However, in contrast to our aforegoing tests with a constant rate sender, overloading the host bypassing system, 36.63 μs latency is comparably low thanks to the TCP congestion control.

To understand the influence of the PCIe topology in more detail, we perform an input rate sweep from 100 Mbit/s to 9.99 Gbit/s on the DUT. The presented results in Fig. 5.6 depict the average latency for a given input rate and test scenario. Further,

Table 5.2 Achievable bidirectional throughput of three concurrent TCP streams with a MTU limit of 1514 bytes in evaluation Senario 3 depending on the PCIe topology

	Avg. goodput (Gbit/s)	Avg. latency (μs)	Loss (%)
Intel Xeon 4110	7.77	36.63	1.84
AMD Epyc 7402	9.28	13.89	0.00
Broadcom PEX 8747	9.29	13.12	0.00

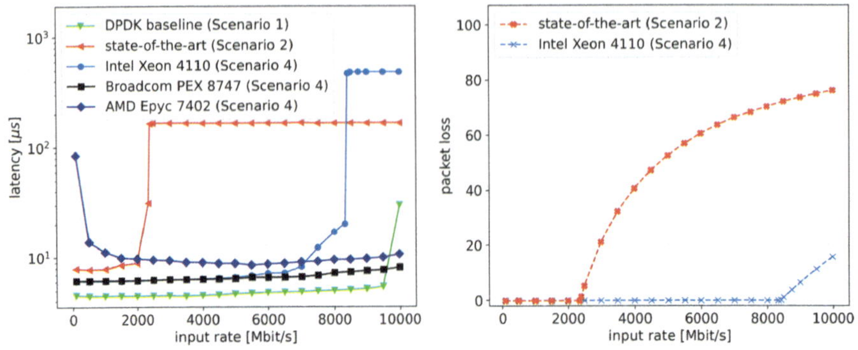

(a) latency (b) packet loss

Fig. 5.6 Packet loss and latency of host bypassing for the different evaluated PCIe topologies depending on the input rate at 300 byte packet size. If no packet loss is denoted, this topology shows up no lost packets for any input rate

if the packet loss is not zero for all rates of a single DUT, the same-colored dotted lines indicate the relative packet loss. Note the logarithmic y-axis scaling.

First, we will discuss the *DPDK baseline* measurement: As stated earlier, this approach does not have any packet loss up to a throughput of 9.99 Gbit/s. However, starting from 5.00 Gbit/s, we observed a slight latency increase. Only at 9.99 Gbit/s, the average latency increases strongly. This latency increase can be explained by the latency spikes, already known from Fig. 5.4, which cannot be retrenched quickly as the link speed of the egress port is very close to the arrival rate. For lower input rates, this packet backlog can be retrenched much faster, wherefore the average latency is significantly lower.

The *state-of-the-art* approach, copying packets via the system's main memory into the hardware accelerator, has a latency of less than 10 μs until a rate of 2 Gbit/s. Increasing the input rate further leads to a significant latency increase up to 168 μs. Further, between 2.35 and 2.4 Gbit/s first packet loss occurs. A further increase of the input rate does not lead to any throughput increase or latency variation.

Utilizing the *Broadcom PEX 8747* PCIe switch causes no packet loss at any input rate. In addition, we observed only a slight latency increase while increasing the load, more precisely 1.95 μs from the lowest to the highest input rate.

The results for the *Intel Xeon 4110* CPU are similar; a latency increase can be observed while increasing the input rate. However, in contrast to the PCIe switch, the system's performance limit is reached before reaching the link speed. Specifically, we observed a significant latency increase between 7 and 8.4 Gbit/s but still no packet loss. First packet loss occurs with a load of 8.42 Gbit/s, and the goodput ratio does not increase further with higher input rates.

The FPGA toolchain provides an integrated logic analyzer, which we utilized for further debugging this performance. We observed that the incoming data write bursts at the PCIe interface are at most 64 bytes long. In the case of the PCIe switch, we observed longer write requests. Thus, we can assume a PCIe write burst fragmentation within the CPU. However, we neither know the reason for this fragmentation nor can we confirm this as the reason for the limited performance. Possibly the discrepancy of the PCIe speeds, specifically Gen2 for the NIC and Gen3 for the FPGA, causes this fragmentation.

Last, we investigate the *AMD Epyc 7402* CPU, which causes zero packet loss with all tested input rates. Contrary to the other two PCIe architectures, this root complex causes a higher latency for very low rates. With an increased input rate until 5.5 Gbit/s, the latency decreases. From this point on, the latency slightly increases, similar to the other architectures. This behavior is comparable to memory caches and might be caused by a routing table cache within the CPU root complex for physical addresses. However, we can neither confirm nor falsify this assumption.

Summarized, we conclude that the PCIe architecture and topology strongly influence the *host bypassing* performance with FPGAs.

5.1.2.4 Tail Pointer Delaying

In Sect. 3.1.4, we introduced the concept of batching rx and tx tailpointer updates to lower the PCIe bus utilization. To evaluate this, we performed three experiments, as shown in Table 5.3. In the first run, the FPGA writes tailpointer updates immediately into the NIC, equating the hitherto discussed approach. The second and third runs describe a tailpointer update only for every 8_{th} or 16_{th} packet, respectively. Latest after 2,500 ns, the tailpointer will be updated to avoid packets being never sent. In the third run with a batch size of 16 packets, this timeout occurs frequently due to the per-packet transmission time of 240 ns (a result of the link speed and packet size).

The batching of tailpointer updates causes an increased latency while sending packets, as the NIC becomes informed later of newly available packets. However, we measured a lower standard deviation in the latency distribution, presumably caused by the lower utilization of NIC, FPGA, and PCIe resources. Further, the NIC can read multiple descriptor ring entries at once.

Table 5.3 Influence of delaying and batching the rx and tx tailpointer updates in Scenario 4. The test load is 300 byte packets at 9.99 Gbit/s

Batch-factor/timeout (ns)	Latency		PCIe transfers (bytes/pkt)	
	Average (μs)	Std. deviation (ns)	Outgoing	Ingoing
1/-	7.86	101.23	324.0	316.0
8/2500	9.15	98.93	317.0	316.0
16/2500	9.43	65.33	316.7	316.0

In addition to the latency, we monitored the ingoing and outgoing PCIe bus transactions within the FPGA. Here, we observe a decrease in the number of bytes sent out of the FPGA, which is caused by the reduced number of tailpointer updates.

If a latency increase is acceptable, we can assess the tailpointer batching mechanism as a viable solution to lower the PCIe bus load. However, only delaying the rx tailpointer updates causes no latency increase as long as sufficient free descriptor ring entries are available.

5.1.2.5 Tx Descriptor Writeback

The utilized NIC and the presented FPGA prototype provide the capability of an optional tx descriptor writeback, as introduced in Sect. 3.1.4. In Fig. 5.7a, we present the measured latency over time for the bidirectional Scenario 4 with and without tx descriptor writeback and two PCIe architectures.

In the case of the NIC and FPGA being attached to the PCIe switch, we cannot determine any performance difference caused by the writeback. The latency is constantly low, and zero packet loss occurs.

However, the behavior for the *Intel Xeon 4110* CPU changes when writeback is enabled. With enabled tx descriptor writeback, 15.79% packet loss occurs, significantly more than the 9.90% for the other case. This can be explained by the additional overhead caused by writing an 8−byte descriptor for every packet into the FPGA.

The advantage of *tx descriptor writeback* is congestion control for sending packets, as every sent packet is confirmed in the tx descriptor ring. By that, overflowing the descriptor ring becomes impossible, e.g., if the enqueueing mechanism of the FPGA is faster than the tx mechanism of the NIC. The red crosses in Fig. 5.7a indicate reordered packets caused by descriptor ring overflows. We define a reordered packet if its first timestamp, i.e., before the DUT, is smaller than the timestamp of the preceding packet behind the DUT. In this test, we observed 176,761 packets fulfilling this reordering criterion, while a total of 35,784,597 packets were processed.

The diagram in Fig. 5.7b depicts the time of 20 consecutive packets staying within the FPGA with no descriptor writeback enabled. The snapshot was created 0.2 ms after the start, where the packet queue was not yet fully built up, in order to create a readable graph. The left side of a horizontal bar represents the ingress time, the right side of the bar when the NIC sends out the packet. One can observe two classes of packets exist, and the latency difference of packets between these classes is around 15 μs, approximately the round-trip-time of the tx descriptor ring with 64 entries, 300−byte packets, and 10 Gbit/s link speed. For example, the Packets 8, 9, and 10 cause a tx ring overflow and are sent earlier than expected. The Packet 11, however, is an "old" packet of the previous descriptor filling round.

To summarize, a *tx descriptor writeback* is only required when the maximum system performance is reached, and packet reordering should be avoided. Yet, the price of descriptor writeback is a lowered performance due to the additional PCIe overhead.

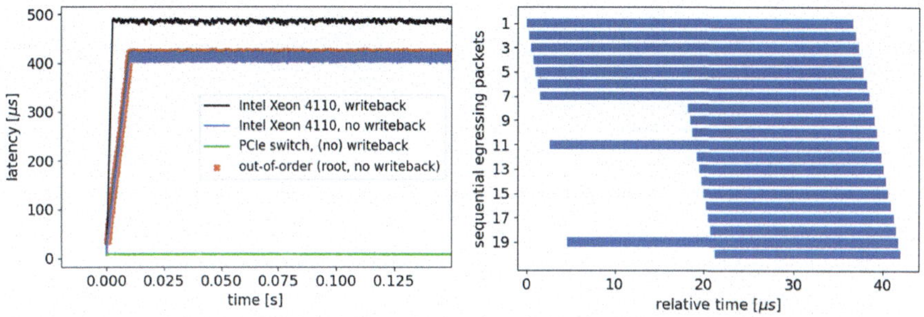

(a) Latency and out-of-order packets over time.

(b) Relative ingress and egress time of 20 consecutive packets 0.2 *ms* after start.

Fig. 5.7 Packet reordering in an overloaded scenario. Tx descriptor writeback is disabled/enabled. The test load is 300 byte UDP packets at 9.99 Gbit/s

5.1.3 GPU-Based Host Bypassing

As an extension of the *host bypassing* approach with FPGAs, we proposed a realization on GPUs in Sect. 3.1.5. Identically to the FPGA approach, packets are received from the GPU via PCIe from the NIC, without any CPU or main memory interaction.

We set up a test environment to evaluate this approach, as shown in Fig. 5.8. Test packets can ingress and egress the DUT only via the NIC, as the GPU provides no network connectivity. The remaining measurement setup, based on the P4STA framework, is unchanged from the previous section.

Specifically, we compare the two scenarios as shown in Fig. 5.9. The first scenario is the *DPDK baseline*, containing only packet I/O to and from the system's main memory. As discussed in the previous section, we assume this scenario as an upper bound for state-of-the-art packet I/O into hardware accelerators.

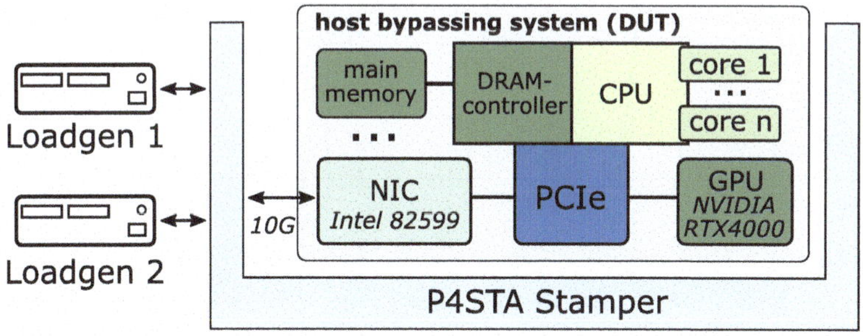

Fig. 5.8 Measurement setup for the host bypassing approach with GPUs building upon the P4STA framework. © 2021 IEEE. Reprinted, with permission, from "Host Bypassing: Let your GPU speak Ethernet" [8]

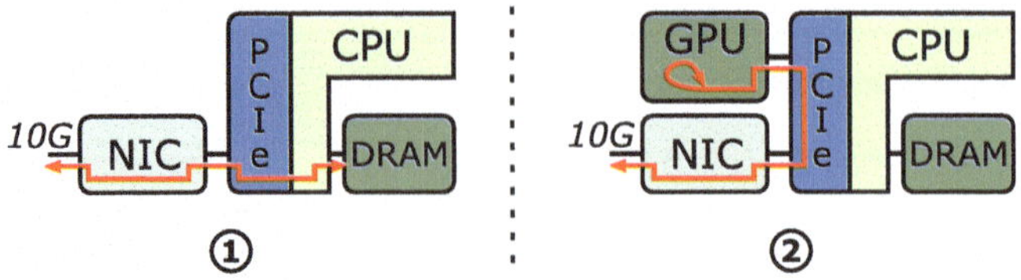

Fig. 5.9 Evaluation scenarios for host bypassing with GPUs. © 2021 IEEE. Reprinted, with permission, from "Host Bypassing: Let your GPU speak Ethernet" [8]

Instead of building a state-of-the-art reference implementation for packet I/O into GPUs, we only compare the GPU *host bypassing* implementation with this upper bound. The second scenario specifies the *host bypassing* approach with GPUs.

In this section, we will discuss the performance of *host bypassing* with GPUs, focusing on latency, jitter, and throughput. Following this, in Sect. 5.1.4, we compare the measurement results of FPGAs and GPUs with each other.

5.1.3.1 General Packet I/O Performance

The plots in Fig. 5.10 show the latency over time for the evaluation Scenarios 1 and 2 with a packet size of 300 bytes and 1000 bytes at 9.99 Gbit/s. The GPU *host bypassing* implementation consists of 8 descriptor rings for receiving and 8 descriptor rings for sending packets. The influence of the number of rings will be discussed in a subsequent section. Analogous to the FPGA evaluation, the NIC and GPU in

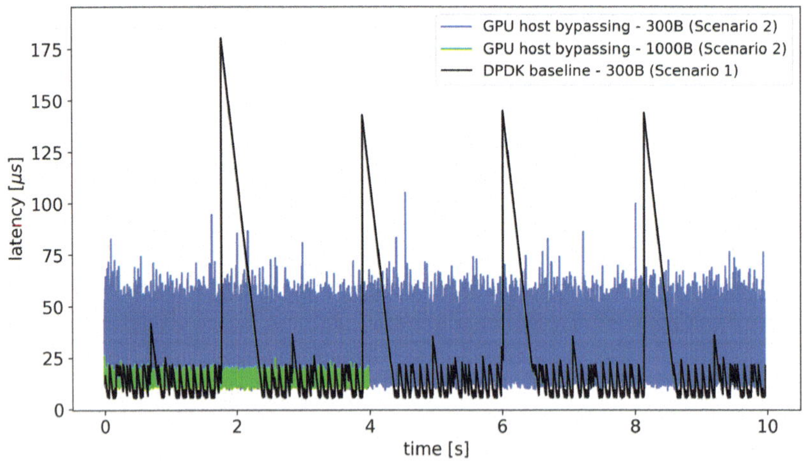

Fig. 5.10 Latency over time for GPU host bypassing (8 rings) with 300-byte and 1000-byte test packets at 9.99 Gbit/s. The green curve is cut-off for readability at 4 s

Scenario 2 are connected to a PCIe switch, and in Scenario 1, the NIC is directly attached to the CPU root complex to be closest to the system's main memory.

Investigating the curve for GPU *host bypassing* with 300−byte packets, we observe significant latency jitter, ranging from ∼8 to ∼ 100 μs, with an average latency of 27.17 μs. With an increased packet size of 1000 bytes, the jitter is significantly lower, and the average latency is 15.28 μs.

Compared to this, the *DPDK baseline* has a low median latency; however, periodic latency peaks disturb the overall system performance. On average, the baseline has a latency of 29.19 μs, which is strongly raised by the latency peaks.

From this experiment, we can derive two insights: First, *host bypassing* with GPUs is susceptible to jitter and a higher packet rate, i.e., smaller packets at constant bitrate, increases this effect. Second, compared to *state-of-the-art* approaches, with a theoretically bounded performance by the *DPDK baseline*, this approach is a viable solution to improve the packet I/O performance.

5.1.3.2 Parallel Packet Processing

In the following, we investigate the influence of parallel packet receiving and sending in the GPU. The GPU implementation of *host bypassing*, presented in Sect. 3.1.5, allows the instantiation of multiple descriptor rings for receiving and sending packets in parallel.

The NIC classifies every incoming packet by computing a hash value based on the packet header fields and assigns it to one of the receive rings. We generated synthetic test traffic for the evaluation, consisting of multiple packets equally distributed over all available hash values. By this, we utilize all descriptor rings uniformly.

The results in Fig. 5.11 are generated with 1000−byte test packets in a non-overloaded scenario. Consequently, zero packet loss occurs in all tests. We observe a slight rise in latency when the number of descriptor rings becomes greater than one. Further, higher variance in latency is visible, i.e., higher latency jitter. Although we cannot observe the internal behavior of the GPU in real-time, we can assume possible causes. The parallel descriptor rings are served by multiple CUDA threads within the GPU, each accessing the same global memory. Probably this synchronous access leads to a disturbance between the threads.

Next, we will investigate the GPU performance depending on the number of descriptor rings and packet size at a data rate of 9.99 Gbit/s. The results are presented in Table 5.4.

First, we can observe a very high latency for all tests with packet loss. The reason for this is the rx buffer of the NIC, already discussed in the preceding section.

Second, zero packet loss occurs for large packet sizes, i.e., 1500 byte, even for only a single descriptor ring. From this, we can identify the packet rate to be the limiting factor. Further, the latency standard deviation is very low for a single descriptor ring and large packets. Therefore, as long as large packets can be guaranteed, a single descriptor ring should be preferred for jitter-critical applications.

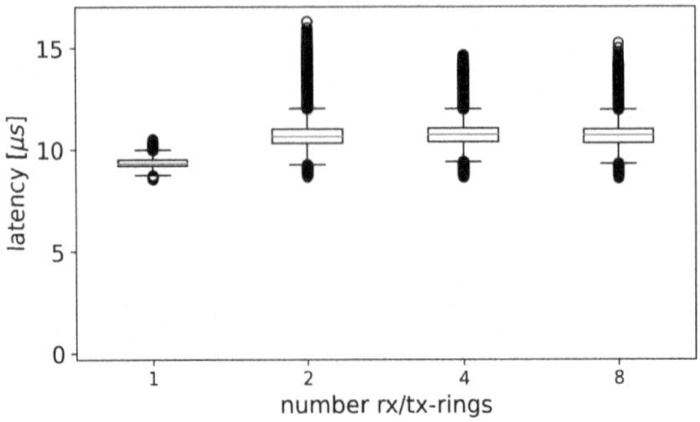

Fig. 5.11 Latency distribution of host bypassing with GPUs depending on the number of rx/tx descriptor rings. 1000-byte test packets at 7.4 Gbit/s

Table 5.4 GPU host bypassing performance characteristics, influenced by the number of rx and tx rings and packet size, at a constant 9.99 Gbit/s UDP test load

rx-/tx-rings	Packet size	Avg. latency (μs)	Latency std. dev.	Packet loss	Reordered packets
1/1	1500	9.76	87.7 ns	0	0
1/1	300	1790	50.3 μs	74.78%	0.01%
2/2	300	1000	87.7 μs	51.64%	49.49%
4/4	300	544.2	99.7 μs	6.45%	42.28%
8/8	300	27.17	7.6 μs	0	38.20%

Third, we observe massive packet reordering for multiple descriptor rings. For this, analogous to the FPGA *host bypassing* evaluation, we captured a timestamp with nanosecond accuracy for every packet before and after the DUT. This reordering can be explained by the asynchronous processing of the rings in the GPU, implicitly allowing packets to overtake other packets. Although packet reordering is an issue for some applications, for many network functions it is irrelevant. In the case of a single descriptor ring and 300-byte packets, the reordering is caused by overflows in the descriptor ring. For 300-byte packets at line rate, zero packet loss occurs starting with 8 descriptor rings.

5.1.3.3 Influence of the PCIe Topology

In Sect. 5.1.2, we have shown that the performance of the *host bypassing* approach with FPGAs significantly depends on the PCIe infrastructure.

We tested the DUT with a constant UDP test load of 300−byte packets and 4 descriptor rings for receiving and sending packets to evaluate the PCIe influence. Four rings are intentionally chosen, as this causes an overloaded scenario for all topologies and the packet loss is a good metric for the achievable throughput. The results are presented in Table 5.5.

Table 5.5 Influence of the PCIe topology on the host bypassing performance. All tests with 300-byte test packets at 9.99 Gbit/s and 4 descriptor rings for rx and tx

	Avg. latency (μs)	Latency std. dev. (μs)	Packet loss (%)
Broadcom PEX 8747	544.18	99.68	6.45
AMD Epyc 7402	564.67	94.75	7.72
Intel Xeon Silver 4110	583.73	98.36	9.82

The packet loss is lowest for the PCIe switch (*Broadcom PEX 8747*) and highest for the *Intel Xeon 4110* CPU. The latency measurements have only a low significance as they are mainly caused by the rx buffer in the NIC, which cannot write the packets faster into the GPU. This buffer has a constant size, so the latency depends inversely on the throughput.

We can summarize that the PCIe topology influences the performance of *host bypassing* with GPUs slightly, but all investigated architectures have shown a similar and good performance.

5.1.4 Comparison of Host Bypassing with FPGAs and GPUs

In the two preceding sections, we discussed the performance characteristics of the *host bypassing* approach with FPGAs and GPUs. Although both realizations rely on the same concept, their performance characteristics are diverse.

First, the **latency** of GPU host bypassing is significantly higher than with FPGAs, as shown in Fig. 5.12. A higher latency leads to a higher backlog in the system, i.e., the number of packets being processed concurrently. Further, we observed a higher **jitter** for all experiments with GPUs. Both effects depend probably on the internal architecture of the GPU, consisting of a DDR5-based global memory on which the software threads operate. In contrast to this, the FPGA realization is highly optimized for this particular application, and therefore packet handling can be much faster. Indeed, both approaches outperform *state-of-the-art* approaches, handing over packets via the system's main memory. The line rate of the NIC, i.e., 10 Gbit/s, can be reached with both approaches.

While the FPGA implementation reached a very high performance with a single descriptor ring for sending and receiving packets, the GPU implementation benefits from **parallelism**. In our tests, we observed an increased achievable throughput when increasing the number of descriptor rings up to 8 for sending and receiving.

The influence of the **PCIe architecture** influences the overall performance. The PCIe switch has shown the best performance for FPGA and GPU *host bypassing*, followed by the PCU root complexes of AMD and Intel CPUs. Surprisingly, these effects were much stronger for the FPGA implementation. Perhaps, this is caused either by the GPU being optimized for common CPUs or by the parallel packet handling of the GPU and by this more simultaneous PCIe bus transfers. The faster

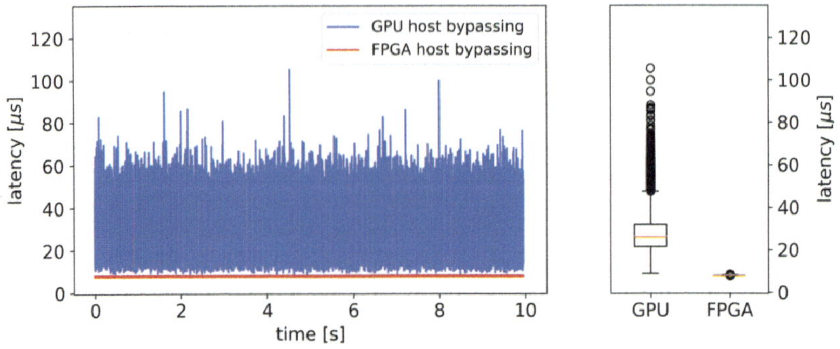

Fig. 5.12 Comparison of host bypassing latency for FPGAs (1 ring) and GPUs (8 rings) at 9.99 Gbit/s and 300-byte packets. No packet loss in both measurements

link technology of the AMD CPU, supporting up to the 4_{th} PCIe generation, should not influence the results, as neither the NIC nor the FPGA/GPU supports this.

Tailpointer batching is an improvement that can be realized only with FPGAs at an acceptable effort. Yet, it can lower the PCIe bus utilization and increase the system's performance along with a slight latency increase.

To summarize, we can state that the *host bypassing* approach is superior to *state-of-the-art* techniques for Packet I/O in PCIe-based hardware accelerators.

5.2 Network Function Offloading on Programmable Hardware

In Sect. 3.2, we introduced flexible Internet service creation concepts using programmable hardware. In the following, we will evaluate the performance of these concepts. First, in Sect. 5.2.1, we discuss the residential Internet service creation with P4-programmable hardware, including switches, smart Network Interface Cards (NICs), and Field Programmable Gate Arrays (FPGAs). Next, in Sect. 5.2.2, we investigate the mobile Internet access use case, focusing on the User Plane Function (UPF).

As these two residential and mobile Internet access creation approaches provide only the subscriber termination but no per-subscriber traffic shaping and QoS enforcement, we suggested an FPGA-based QoS co-processor. The prototypical implementation of this co-processor will be evaluated in Sect. 5.2.3 to prove the underlying concepts.

5.2.1 *Residential Internet Access Termination*

The requirements for residential Internet service creation, i.e., a Broadband Network Gateway (BNG) system, are quite challenging. Especially, fulfilling the requirements regarding the number of parallel subscriber sessions, throughput, and application-specific network protocols is challenging (compare Sect. 2.2). In the following, we will investigate the performance criteria of the proposed prototype, i.e., latency, throughput, and the number of subscribers.

5.2.1.1 Evaluation Setup

In total, we investigate three different P4-programmable targets, all realizing the desired BNG functionality, and a software-based baseline implementation for pure packet forwarding:

- **P4-Tofino:** The *P4-Tofino* platform is a P4-programmable switch manufactured by Intel (formerly Barefoot Networks) and provides up to 64 *x* 100 Gbit/s Ethernet ports.
- **P4-NetFPGA:** This P4 target offers 4 *x* 10 Gbit/s ports and can be programmed by a P4-compiler, integrated into the common FPGA development workflow [7].
- **P4-SmartNIC:** The *Netronome NFP-4000* is a Network Processing Unit (NPU) that allows programming its data plane with the P4 programming language. It offers up to 2 *x* 40 Gbit/s Ethernet connectivity or up to 8 *x* 10 Gbit/s breakout Ethernet ports.
- **Software-based baseline (Linux Kernel):** This work assumes the superiority of hardware-accelerated Internet service creation over pure software approaches. However, a software reference implementation, e.g., *Accel-PPP*[1], may not offer the maximum achievable performance. Therefore, we compared the investigated BNG implementations with the native packet forwarding of the Linux kernel as the upper bound for kernel space network functions.

The general evaluation measurement setup is shown in Fig. 5.13, building upon the P4STA framework introduced in Chap. 4. The three P4-targets and the software

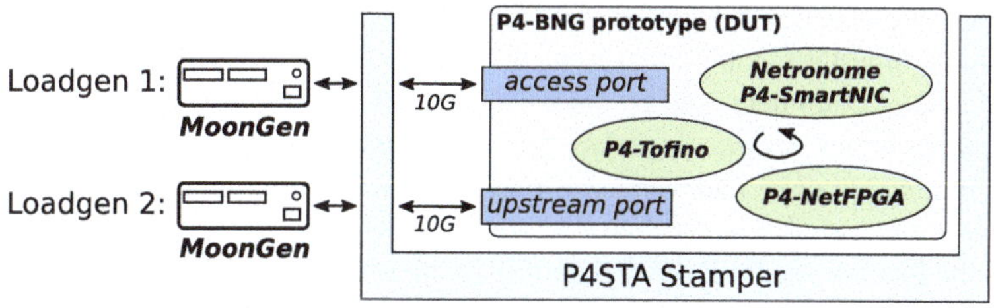

Fig. 5.13 Evaluation scenario for the P4-BNG. The three implementations can be exchanged. All implementations are attached via a 10 Gbit/s Ethernet link

baseline can be exchanged in this setup, each being connected via two 10 Gbit/s Ethernet links. This link speed is chosen for fairness reasons, as Ethernet systems with a higher link speed typically have a lower base latency due to their store and forward behavior. For example, 1500 byte packet requires 1.2 μs to be transmitted on a 10 Gbit/s link, but only 120 ns at 100 Gbit/s link speed. This implies that even a lower latency can be achieved. For this, we refer to Sect. 5.2.2, where we present evaluation results for mobile Internet access termination on the P4-programmable Tofino platform, operating at 100 Gbit/s.

While the P4STA framework is used for packet loss and latency measurements, the test traffic generation is done with the *MoonGen* load generator [5]. *MoonGen* allows generating the custom network protocol stacks, i.e., PPPoE encapsulated subscriber packets, at high rates with various session configurations. Further, a receive thread in *MoonGen* can verify packets on the access port. To summarize, we benefit from the disaggregation flexibility of P4STA, allowing the usage of a highly flexible software load generator.

To increase the packet rate, we lowered the packet size to 532 bytes on the upstream port and 562 bytes at the access port. The additionally added headers for PPPoE-based subscriber encapsulation cause the size difference. With this smaller packet size, the differences between the P4 targets become more visible. Note that all presented latency measurements are corrected by a constant offset for the P4STA setup and cable delays, i.e., the shown latency numbers are caused only by the tested device.

5.2.1.2 P4 Language Limitations

One main goal of the P4 programming language is platform independence, i.e., it should be possible to compile the BNG data plane program for the different platforms. However, we observed multiple restrictions.

First, the P4 language (compare Sect. 2.3), exists in two versions: $P4_{14}$ and $P4_{16}$. These two languages express the same logical constructs but are not compatible with each other. Both languages were about equally common at the starting time of this work. The P4-NetFPGA, building upon the Xilinx *SDNet* compiler, offers only $P4_{16}$ support. The Netronome SmartNIC and the Intel Tofino provide support for both languages. However, the $P4_{16}$ support was very restricted at the starting time of this work. Note that in the meantime, the $P4_{16}$ support for the Intel Tofino is significantly enhanced. We tested and used both versions of the Intel Tofino in this work; for the Netronome SmartNIC, only $P4_{14}$ is used.

The *SDNet* compiler, used by the P4-NetFPGA project, allows neither *longest prefix match* tables nor *parser_value_sets*. However, both are used in our BNG reference implementation, and we circumvented them. Note that this is not a limitation of FPGAs; instead, the early-stage P4 compiler does not support these features.

For the Netronome NFP-4000 SmartNICs, we observed multiple minor issues at compile time. This may be caused as the primary programming language for this hardware is *micro-C*, a vendor-specific C dialect. For example, we observed table size limits that do not correspond to the configured sizes.

The Intel Tofino platform is built for the P4-programming language and has the best support for this language. However, if a very high resource utilization or special functionality is desired, the source code must be adopted as well. For these optimizations, detailed knowledge of the hardware pipeline is required.

The P4 language specification, a reference compiler, and the *bmv2* software switch reference implementation are available open-source. It is noticeable that commercial tools are utilized for all of the three investigated platforms. Further, confidentially agreements are partially mandatory to access compilers and documentation, not providing an ideal basis for a quick establishment in industry and academia.

Besides the Intel Tofino platform, the impression is created that the utilized P4 compilers are not yet fully-grown. However, the FPGA vendor Xilinx recently announced a successor of the tested P4 compiler, claiming better feature support and commercial suitability [11]. In general, we can confirm that the P4 language is a promising step towards platform independence. As of today, minor adaptions to the source code are required to enable a compilation on multiple platforms.

5.2.1.3 General Performance Characteristics

First, we evaluate the latency and throughput of the different hardware platforms, realizing the BNG functionality. In the following, we measure the latency and packet loss of the Device Under Test (DUT). Note that zero packet loss at a constant input rate is the demonstration that (at least) this throughput can be handled by the DUT. The "point of failure," i.e., the maximum throughput, can be determined by performing multiple measurement runs at multiple input rates.

Figure 5.14 shows the average latency for the different P4-BNG realizations depending on the input rate. In addition to this, Table 5.6 presents the latency standard deviation for selected input rates of the same measurement scenario. The standard deviation is a metric for jitter and indicates how the latency values of different packets vary around the average value. By this, we can investigate how deterministic the DUT behavior is.

The **P4-Tofino** latency curve has a constant and low latency of 1.17 μs, and no influence of the varying input rate can be observed. The measured latency standard deviation is in the range of the measurement error and can be assumed to be almost zero. As this switch builds upon a P4-programmable ASIC with a total capacity of 6.5 Tbit/s and 100 Gbit/s per port, this behavior is not surprising. The packets traverse without any interference through the switch pipeline in constant time. As the switch is a store-and-forward Ethernet device, the packet is stored at least once internally, which already requires 425 ns at the configured link speed and packet size.

Similarly, the **P4-NetFPGA** has a constant latency, independent of the input rate. However, this latency is higher, i.e., 5.08 μs on average. Executing exactly the same digital logic on FPGAs with a lower clock frequency, e.g., the described P4 pipeline, would cause significantly higher latency. However, these results are still very good compared to the *Linux baseline*.

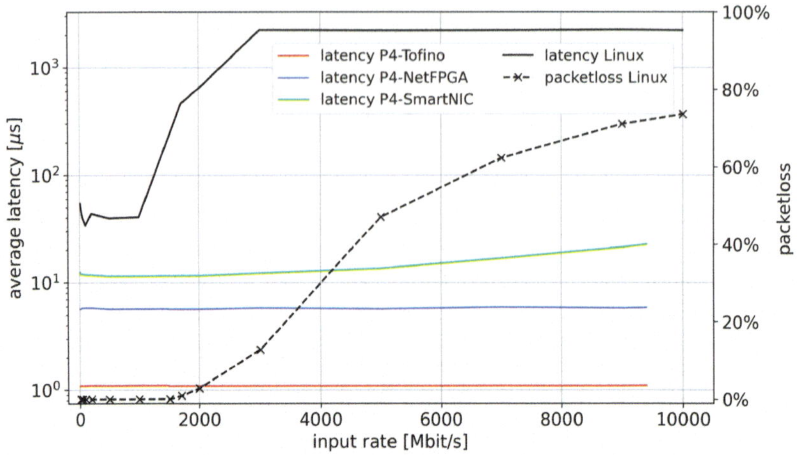

Fig. 5.14 Average latency of the tested BNG system in the downstream direction, depending on the input rate. Packet loss exists only for the Linux kernel packet forwarder. Each run is 10 *s* long. An input rate of 9.4 Gbit/s corresponds to ~10 Gbit/s at the egress port due the additional added headers

Table 5.6 Latency standard deviation of the P4-BNG implementation for selected input rates of the scenario in Fig. 5.14

Input rate:	P4-BNG			Baseline
	P4-Tofino (ns)	P4-NetFPGA (ns)	P4-SmartNIC (μs)	Linux (μs)
100 Mbit/s	<10	327	1.7	13.2
1, 000 Mbit/s	<10	211	0.7	19.8
9, 000 Mbit/s	<10	263	1.6	157.9

The **P4-SmartNIC** has the highest latency of the three investigated platforms, about 22 μs at 9.4 Gbit/s. Further, we observe a latency increase with an increasing input rate. In contrast to the two other P4 targets, this hardware builds upon a many-core processor for parallel packet processing. Therefore, a higher load can cause internal micro congestion, leading to the observed latency increase. Surprisingly, the latency standard deviation decreases for higher input rates, probably because of a more constant internal resource utilization. However, we cannot explain the causes of this behavior without knowing the platform's hardware architecture details.

In comparison to this, the performance values of the **Linux baseline** differ significantly. For all input rates, the latency is significantly higher. Starting with a rate of ~1.5 Gbit/s, the first packet loss occurs, and the latency increases drastically. This latency increase can be explained by packet queues built up in the server system or ingoing Network Interface Card (NIC), as the software cannot process them as fast as they arrive. But even before this *point of failure*, we observed non-deterministic behavior. Especially for very low rates, e.g., 100 Mbit/s, higher latency is measured than for higher rates. This may be caused by the processor cache, which benefits from a higher packet rate.

In general, we can conclude that all three hardware approaches have a significantly better performance than the software baseline in terms of latency, packet loss, and throughput. Note that this Linux baseline performs no packet processing at all and would thus be an advantage compared to the P4-platforms.

5.2.1.4 Number of Subscriber Influence

Due to the nature of Internet service creation, many subscribers should be terminated on a single access edge, e.g., up to 35,000 for a BNG system [12].

Table 5.7 shows the average latency depending on the number of installed subscriber sessions. In all measurements, zero packet loss occurs. No results were shown for the *Linux baseline*, as this is not able to terminate subscriber sessions.

The *P4-Tofino* and *P4-NetFPGA* are not sensitive to the number of parallel sessions, which is not surprising due to the foregoing evaluation results and their deterministic internal logic.

For the *P4-SmartNIC*, we observed a slight variation in latency. This may be caused by the internal load balancer, distributing incoming packets over multiple processor cores. However, further investigations would require detailed knowledge of the proprietary hardware and compiler.

Further, for unknown reasons, more than 4,000 installed subscribers on the *P4-SmartNIC* lead to unexpected packet loss. We assume this to be caused by a control plane bug, not installing flow rules going beyond 4,000 subscribers.

All in all, the behavior of all three P4 platforms is almost independent of the number of subscribers. As expected, the hardware behaves without difficulty until its capacity limit. After this, all further subscribers are ignored. The *P4-SmartNIC* compiler either misbehaves or does not refuse a P4-program exceeding the hardware capacities. According to the datasheet, the available table memory of the *P4-SmartNIC* is higher; thus, we assume compiler misbehavior.

Table 5.7 P4-platform dependent average latency of the P4-BNG for a variable number of subscribers and 9.4 Gbit/s downstream traffic. Input packet size: 532 byte

	#Subscribers					Std. dev. (#S = 4000)
	32 (μs)	256 (μs)	1024(μs)	2048 (μs)	4000 (μs)	
P4-Tofino	1.17	1.17	1.17	1.17	1.17	1.54 ns
P4-NetFPGA	5.08	5.08	5.08	5.08	5.08	242 ns
P4-SmartNIC	22.12	22.14	22.19	22.03	22.06	2.6 μs

5.2.1.5 Resource Utilization

Tables 5.8 and 5.9 show the resource utilization of the BNG prototype on the *P4-NetFPGA* and *P4-Tofino*, respectively. As the used *P4-SmartNIC* builds upon a many-core processor, a NPU executing program code, no resource utilization can be discussed.

FPGAs consist of many different resources that can be configured by the synthesis tool to achieve the desired behavior. However, only the resources Lookup Table (LUT) and Block Random Access Memory (BRAM) are highly utilized by the BNG implementation. LUTs are mainly used for realizing boolean logic, while BRAMs are small Static Random Access Memory (SRAM)-based memory cells, e.g., used for realizing packet forwarding tables.

The utilization values in Table 5.8 show these two resources. The *total* value describes the utilization of all logic on the *P4-NetFPGA*, including Ethernet logic and PCIe control plane integration. The *P4-datapath* value includes only the P4-programmable match-action pipeline. Even though the available resources are not yet fully utilized. A higher utilization is not possible as the synthesis tool must place them on the FPGA, fulfilling strict timing guarantees, i.e., how long an electrical signal takes from A to B. Therefore, ~4000 subscribers are already the maximum on this platform, depending on the required functionality.

SRAM memory and Ternary Content-Addressable Memory (TCAM) are the critical resources in the *P4-Tofino*. In Table 5.9, the utilization for 4096 and 8192 subscribers is shown. Similar to the *P4-NetFPGA*, a utilization of 100% is not realistic due to dependency conflicts of sequential P4 constructs. However, a higher utilization than for FPGAs, including the *P4-NetFPGA*, is possible. The logic is distributed over nine consecutive pipeline stages for dependency reasons, each having a constant amount of memory and logic blocks.

Table 5.8 Resource utilization of the P4-NetFPGA platform, realizing BNG functionality for 4096 subscribers

	Total	P4-datapath	Available
LUT	202,155 (47%)	163,013 (38%)	433,200
BRAM	1,074.5 (73%)	952 (65%)	1470

Table 5.9 Resource utilization of the P4-programmable Intel Tofino, realizing BNG functionality for 4096/8192 subscribers

	4096 subscribers	8192 subscribers
SRAM (%)	12.81	15.94
TCAM (%)	15.97	15.97
#pipeline stages	9/12	9/12

The increase of subscribers by a factor of two has only a minor impact on the resource utilization, and even higher subscriber numbers can be achieved. The exact number of achievable subscribers depends on the particular configuration. Thus, the 35,000 subscribers stated at the beginning can be achieved, depending on the exact functionality.

5.2.2 Mobile Internet Access Termination

In extension to the residential Internet access termination, in Sect. 3.2.3, we proposed a concept for hardware accelerated subscriber termination in mobile 5G access networks, i.e., the UPF. In this section, we will discuss the performance characteristics of P4-based UPF realizations compared to state-of-the-art kernel space and user space implementations.

Figure 5.15 depicts the evaluation setup, utilizing the previously introduced P4STA framework to evaluate the Device Under Test (DUT). In the case of the P4-based implementation, the UPF is attached via two 100 Gbit/s Ethernet links to the P4STA Stamper. The kernel space and user space realizations of the UPF are attached via two 40 Gbit/s Ethernet links. In addition to these two data plane ports, we developed a simple emulator of the 5G Session Management Function (SMF), installing flow rules in the UPF. By this, isolated testing of the data plane network function becomes possible without setting up a complete 5G network, e.g., no 5G authentication of real User Equipment (UE) is required.

The test packets consist of equally-sized 1000−byte packets shaped to a constant input rate. Further, the *P4STA Stamper* duplicates all incoming packets by a constant factor to achieve packet rates up to 100 Gbit/s. By this, a replayed packet trace of packets on the N3 interface with a rate of 1 Gbit/s and multiplied by a factor of 100 causes a load of 100 Gbit/s on the DUT. After passing through the DUT, all duplicated packets are sorted out, and only the initial load packets are sent to the packet verification.

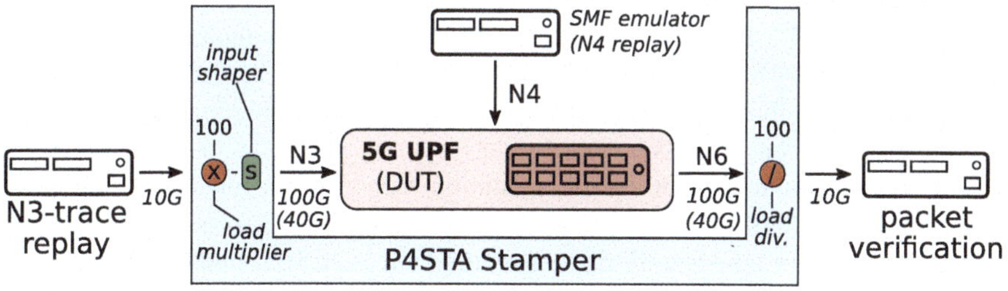

Fig. 5.15 Measurement setup to investigate different UPF implementations. The input load is multiplied by up to 100, enabling higher packet rates

Besides the evaluations in this section, the proposed UPF implementation was tested in an end-to-end testbed, as described in Sect. A.2.

In the next sections, we investigate the following three implementations in detail:

- **Kernel Space UPF (KS):** As state-of-the-art kernel space realization, we utilize the reference UPF implementation of the *free5gc* open-source project [6]. For this, a specific kernel version is required, and a kernel module must be compiled and loaded for GTP encapsulation.
- **User Space UPF (US):** To assess user space subscriber termination, we chose an existing open-source project, namely *dpdk_gtp_gateway* [3]. This implementation builds upon the DPDK framework and requires capable Network Interface Cards (NICs). Further, the NIC is decoupled from the operating system network stack, a typical procedure for user space network functions. This implementation does not support the 5G QoS identifiers, which we added in the source code to provide the same behavior as the other two implementations.
- **P4-based UPF (P4):** The evaluated P4-based UPF implementation is presented as part of this work in Sect. 3.2.3, operating with no QoS co-processor.

For the kernel and user space realization, we utilize a commodity server with two Xeon E5-2670v3 CPUs, 256 GB DDR4 system memory, and two Intel XL710 NICs. The operating system is Ubuntu 18.04 with the kernel version *5.0.0-23*. We disabled CPU clock frequency throttling to avoid non-deterministic disturbance by this. Further, no background tasks are executed on the servers in parallel.

5.2.2.1 Performance Characteristics of a Kernel Space UPF

First, we investigate the characteristics of the kernel space UPF implementation. Figure 5.16 depicts the performance characteristics of the kernel space (KS) implementation. For input rates with a step-width of 500 Mbit/s and additional measurement points around the "point of failure," the average latency and packet loss are shown.

First, while increasing the input rate from 100 to 500 Mbit/s, we observe a latency decrease from 62 to 23 μs. This behavior is already known from the *Linux baseline*, discussed in Sect. 5.2.1. From there on, the latency increases monotonously with the input rate. At 2,450 Mbit/s, the first packet loss occurs, and the latency increases. From this point on, the throughput remains constant, and a further input rate increase causes only higher packet loss.

Note that the authors of this implementation focused on a functional reference implementation and did not aim for the highest achievable performance. However, the general behavior agrees with the observations in Sect. 5.2.1 and in related work for kernel space network functions.

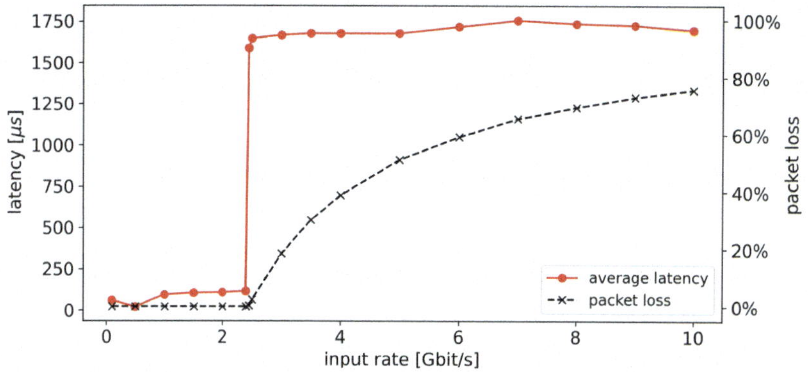

Fig. 5.16 Latency and packet loss of the kernel space UPF implementation, which is provided as part of the *free5gc* open-source project [6]

5.2.2.2 Hardware Acceleration Comparison

Extending the evaluation of the kernel space UPF, we investigated the user space (US) implementation similarly. Up to an input rate of 22.5 Gbit/s, i.e., a rate of $2.8 \cdot 10^6$ packets per second, zero packet loss occurs. From there on, the latency increases, and packet loss occurs. Further, the average latency is significantly lower than the kernel space UPF.

Besides these two software implementations, we evaluated our P4-based UPF prototype in the same way. In Fig. 5.17, the latency distribution of these three different implementations is shown for selected input rates. Note that all shown measurement runs have zero packet loss, which means the DUT is not overloaded.

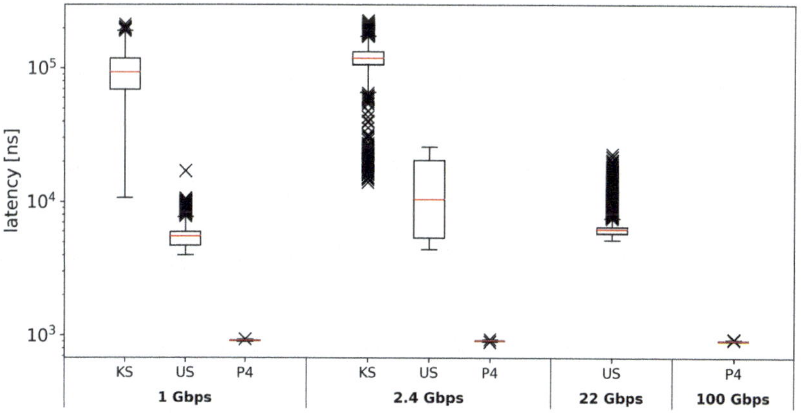

Fig. 5.17 Latency distribution of the kernel space (KS), user space (US), and P4-based (P4) UPF implementations at selected input rates

The **P4**-implementation of the UPF has a constant low latency of ~739 ns on average, independent of the load. Further, almost no jitter is detectable, and zero packet loss occurs, up to 100 Gbit/s. In contrast to this, the P4-BNG has an average latency of 1.17 μs for smaller 532−byte packets (compare Sect. 5.2.1). As discussed before, this difference is mainly caused by the different Ethernet link speeds, i.e., 10 and 100 Gbit/s, and the different P4 programs have almost no impact on the actual latency.

The latency of the user space (**US**) UPF is around one order of magnitude higher than the P4-implementation. It is noticeable that the latency distribution at 2.4 Gbit/s is worse than at 22 Gbit/s. The reasons for this are unknown. As stated before, above 22 Gbit/s packet loss occurs, and the latency increases strongly.

For the kernel space (**KS**) implementation, only results for 1 and 2.4 Gbit/s are shown, as higher input rates would cause packet loss and significantly higher latency (compare Fig. 5.16). The observed latency distribution is again around one order of magnitude higher than for the *user space* UPF and has a high jitter. This corresponds to the known and expected benefit of user space poll-mode drivers.

Note that all implementations may suffer under non-optimal implementations, and the exact performance numbers can vary from implementation to implementation. However, the general finding of our investigations is unmistakable: Between kernel space and user space network functions, significant differences regarding their performance exist. By using programmable hardware, i.e., P4-programmable switches, further performance improvements can be achieved in terms of throughput, latency, and jitter. Therefore the utilization of programmable hardware is mandatory to achieve the ambitious goals of future mobile networks, i.e., low and deterministic latency, high throughput, zero packet loss, and real-time eligibility. Our presented implementation of a 5G UPF has shown the highest performance while providing the required functionality. The proposed concepts can and should be recognized as a trendsetter for next-generation mobile subscriber termination.

5.2.3 QoS-Aware Traffic Shaping in FPGAs

In this section, we evaluate the proposed FPGA-based QoS co-processor, as introduced in Sect. 3.2.4. For this, we focus on the main design requirements, i.e., massive parallel packet queueing for subscriber separation, high throughput, acceptable latency, and zero unexpected packet loss. Note that this generic concept of FPGA-based traffic shaping can be applied for both Internet access scenarios: residential and mobile subscriber termination.

5.2.3.1 Evaluation Setup

To investigate these performance characteristics, we set up an evaluation environment, as shown in Fig. 5.18. Similar to the UPF evaluation in the preceding section,

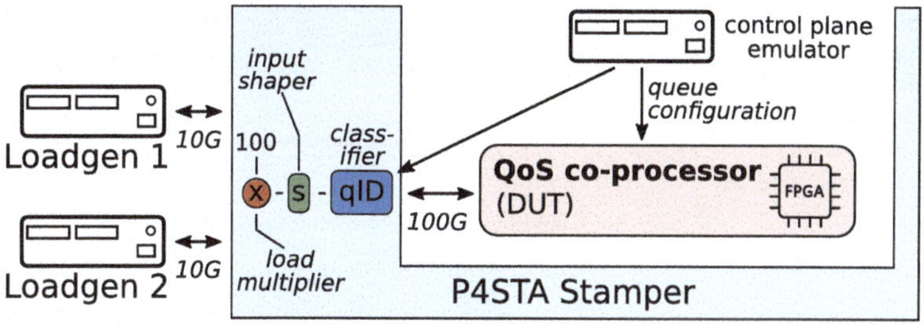

Fig. 5.18 Measurement setup for the FPGA-based QoS co-processor. The P4STA framework is extended by a classifier module, tagging the packets with queue IDs required by the FPGA

the *P4STA* framework is used for packet loss and latency measurements. Further, a load multiplier and input rate shaper are used to generate a fine-shaped test load in the range from a few hundred *Mbit/s* up to 100 Gbit/s.

Further, the *P4STA* framework is used to tag packets with a queue ID *(qID)* (compare Fig. 3.18). For this, a *control plane emulator* configures the classifier within the *P4STA Stamper* and the QoS co-processor with corresponding information. For example, a load generator 5-tuple UDP flow is classified by the *Stamper* device and assigned to the queue ID 42, and a traffic shaping rule for the same queue ID is installed in the QoS co-processor.

As we chose a "sidewise FPGA integration" in Sect. 3.2.4, the packets are ingressing and egressing the QoS co-processor on the same port.

5.2.3.2 General Packet Forwarding Behavior

First, we investigate the maximum throughput of the FPGA-based prototype. The *No-Scheduler* implementation accepts every packet in any queue to be sent immediately, enabling maximum throughput tests. As input load, 1474 byte packets are sent into the FPGA with a rate of 99.9 Gbit/s, enqueued, and as soon as possible dequeued. In this experiment, the packet traverses the complete data path of the FPGA internal design, including the external memory and the internal processing stages (compare Fig. 3.17 of the design section). We utilize two external DRAM-based DDR4 memories in this experiment. In a following section, the impact of using two instead of one external DDR4 memory is investigated in detail.

In Fig. 5.19, the **latency** over time and its distribution are plotted for two different queue numbers, a parameter of the proposed queue design. For 2048 *queues*, the latency varies between 2 and 3 μs. For 262144 *queues*, the lower line of the latency is unchanged, but the maximum latency is ~40 μs. This behavior is caused by the FPGA internal scheduler, following a round-robin mechanism. In each clock cycle of the FPGA, 32 queues can be checked in parallel for packets. The proposed FPGA design operates at a clock frequency of 220 MHz, i.e., ~4.5 ns per clock cycle. With the known number of queues, we can deduce that the scheduler checks every

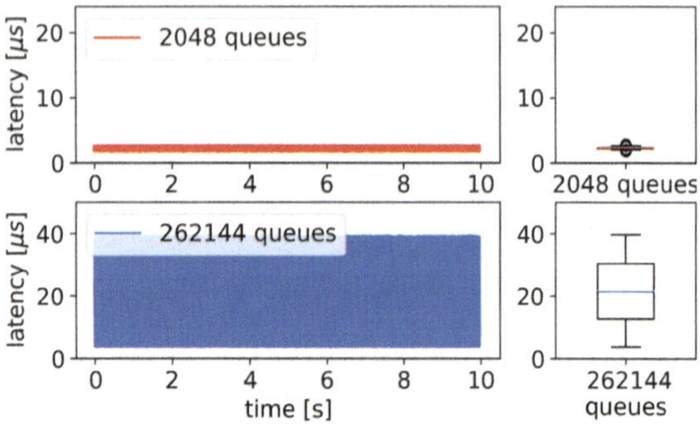

Fig. 5.19 Measured end-to-end latency of the FPGA queueing system with the *No-Scheduler*, performing no traffic shaping. Note the different y-axis scaling

$262144/32 = 8192$ clock cycles the same queue, assuming no other packets are sent in between. This results in a theoretical scheduler round-trip time of 37.23 μs and corresponds with the shown measurement values. In every cycle, multiple packets can be dequeued from a single queue.

Based on these results, we can conclude that the base latency is almost neglectable, and only an increased number of queues causes higher latency.

The achievable **throughput** of the evaluated QoS co-processor is 99.9 Gbit/s, as we measured zero packet loss even at this high input rate. At an input rate of 100 Gbit/s, we observed very little packet loss caused by micro congestion in the FPGA-internal pipeline. However, a 100% link utilization in real computer networks is not possible without packet loss for multiple reasons and is not desirable. Note that this experiment was performed with two external DDR4 memories for packet storing.

5.2.3.3 Fine-Grained Traffic Shaping

In the following, we discuss the working behavior of the token-bucket rate limiters within the scheduler. The results are similar for both schedulers, relying on the same token-bucket implementation, and thus, only the results for the *Simple-Scheduler* are presented in the following. The test load is configured to constant-sized 1000-byte UDP packets, sent into the FPGA with a rate of ~500 Mbit/s. All packets are classified and assigned to the same packet queue in the FPGA. The rate limiter of this queue is set to a rate of ~100 Mbit/s.

In this experiment, only the output of the FPGA is of interest for validating the token-bucket shaper. Observing the packet loss would only show very high packet loss due to the input and output rate mismatch. However, this rate mismatch is required to ensure that the FPGA queue never becomes empty. Further, the latency of the packets is assumed to be very high, i.e., corresponding to the taildrop limit of the queue.

The output behavior of the FPGA, and by this, the traffic shaper, can be investigated best by analyzing the timestamp series of all scheduled packets behind the FPGA.

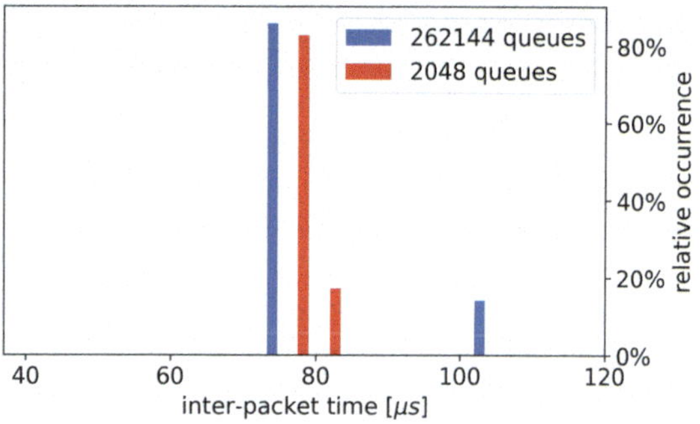

Fig. 5.20 Distribution of the inter-packet scheduling time of the *Simple-Scheduler*, shaping the 1000 byte packets to a constant rate of 100 Mbit/s

The P4STA framework captures these timestamps. In case of varying packet sizes, the size must be captured for each packet in addition.

As every packet is captured, the *inter-packet time* can be computed by subtracting the timestamps from the current and the preceding packet. We can compute the theoretical desired *inter-packet time* for the given numbers as follows:

$$\frac{1000\,\text{byte}}{100\,\text{Mbit/s}} \cdot \frac{8\,\text{bit}}{1\,\text{byte}} = 80\,\mu\text{s}$$

The distribution of the *inter-packet times* is shown in Fig. 5.20 for two different numbers of queues in the FPGA. The start and end phases are cut off as there the input load and queue level did not yet reach the desired steady state. We observe two bars in the plot for each number of queues, one below and one above the expected 80 μs of a theoretical shaper.

The round-robin behavior of the scheduler can explain this behavior. In the scenario with 262, 144 *queues*, the scheduler requires 37.23 μs for one traversal of all queues. Therefore, packets can be sent only every $n \cdot 37.23\,\mu\text{s}$, i.e., after 74.46 μs or 111.69 μs. The token bucket is filled constantly with new tokens, corresponding to a rate of 100 Mbit/s, and the packets are sent either after 74.46 μs or after 111.69 μs. Still the rate limiter enforces a bit rate of 100 Mbit/s on average.

Even though the variance in the inter-packet time seems to be very high, this result is excellent. We can state that this scheduler never sends more than one packet at once, which means that no microbursts occur. As discussed in related work, data center switches with only a few queues compared to the presented prototype tend to send packets in bulks of two or more packets [4]. This is confirmed by our measurements, presented in Sect. 4.3.2, comparing the accuracy of software and hardware-based traffic shaping. In this experiments, performed on an Intel Tofino, we observed microbursts of 8 packets at once at a 100 Gbit/s Ethernet link, shaped to a rate of 100 Mbit/s. Note that this hardware offers only a couple of queues per egress port, which is not comparable with the discussed FPGA prototype in this section.

A token bucket algorithm generally underlies a tradeoff between maximum achievable rate, burstiness, and rate configuration granularity. The correlation of these dimensions is deduced in Sect. A.4. In the presented evaluation measurement, the token bucket algorithm has the following parameter: Every 1024 clock cycles, i.e., every 4.65 μs, new tokens are assigned to the token bucket. As shown in the appendix, microbursts would be unavoidable if this value is higher than the estimated inter-packet time. The token update time of 4.65 μs implies that the shaper has a configuration accuracy of 1.7 Mbit/s. Note that the accuracy can be improved by increasing the token-bucket update time. However, we prioritized a burst-free behavior for even higher rates.

We can summarize that the token-bucket traffic shaper is able to perform accurate traffic shaping and rate enforcing. The observed behavior of the QoS co-processor matches with the expected $M/D/1/B_{max}/\infty/FIFO$ *Kendal* queueing system (compare Sect. 2.4). Even though the exact requirements regarding microbursts in Internet access networks are not specified, we can conclude that this is not a problem, as only a single packet is always scheduled at once. No microbursts occur at all while enforcing the desired rate. Thus, the shaper behavior is sufficient for the use case of Internet access creation.

5.2.3.4 External Memory Influence

We discussed in the design chapter different approaches to realize the external memory for packet storing: First, either a single or two DDR4 memories can be used. DDR4 is a memory technology relying on Dynamic Random Access Memory (DRAM). DRAM allows a very high memory capacity but has comparably high and non-deterministic access time. Second, to overcome this limitation, an optional internal memory, realized with Static Random Access Memory (SRAM), can be instantiated. In this section, the impact of these two configuration decisions will be discussed.

In Fig. 5.21, the average latency of the FPGA design with a single DDR4 memory is shown. The *No-Scheduler* is used to measure the basic packet forwarding behavior

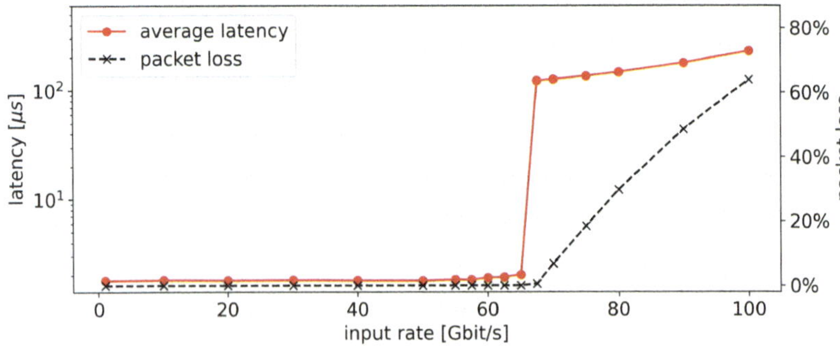

Fig. 5.21 Average latency and packet loss of the FPGA-based QoS co-processor with a *single* DDR4 memory and the *No-Scheduler* depending on the input rate, from 1 Gbit/s to 99.9 Gbit/s

without any congestion. Up to a rate of 65 Gbit/s, zero packet loss is detectable. Further, the latency increases only slightly by ~200 ns in this rate range. An input rate increase causes only higher packet loss from this point on. Surprisingly, the packet loss is disproportionate. For an input rate of 100 Gbit/s, 63.89% packet loss occurs. This packet loss would imply that the maximum throughput is around $(1 - 0.64) \cdot 100 = 36$ Gbit/s, which is lower than the maximum loss-free rate of 65 Gbit/s. An FPGA internal memory data bus analysis shows that the higher input rate causes more packet write-requests on the DDR4 memory, lowering the read performance for sending packets.

Compared to related work and data sheets, stating that the memory bandwidth of DDR4 memory is higher than 100 Gbit/s, the measured results seem to lower than expected. However, in our prototype, the memory interface writes *and* reads the data of each packet. Theoretically, by this, the bandwidth is lowered by a factor of two. Further, with the change of operation mode, i.e., reading and writing data simultaneously, the achievable memory bandwidth is even lower [14]. Therefore, we can conclude that the measured results are in the range of maximum achievable performance. If a throughput higher than ~65 Gbit/s should be achieved, two external DDR4 memories are required. The discussed results in the previous section show that with two memories, zero packet loss up to 100 Gbit/s is achievable (compare Fig. 5.19).

In addition to the external DDR4 memory, an internal SRAM memory can be used for small packets. It is common knowledge that SRAM memory has a deterministic behavior and a lower access latency. Following, we investigate the influence of this theoretical advantage. Figure 5.22 depicts the latency distribution of probe packets for different packet sizes. Measurements for the SRAM memory are only shown for 250 and 500 byte packets due to the memory alignment of this memory, i.e., allowing only packets up to 512 byte.

First, we observe a higher latency variance for the DDR4 memory compared to the SRAM. Especially, the outliers of the DDR4 memory are conspicuous. They confirm our suspicion that DDR4 memory is not well suited for network applications with

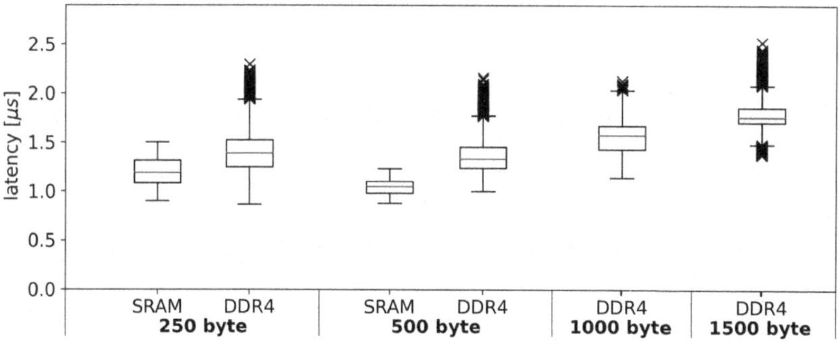

Fig. 5.22 Latency distribution of the QoS co-processor, depending on the memory technology and packet size

very low and deterministic latency requirements. Even so, we cannot determine the cause of these outliers. They are likely caused by memory refresh cycles running parallel to the packet handling.

Second, the latency distribution of SRAM memory for 500-byte packets is better than for 250-byte packets. The constant input data rate causes this, i.e., the packet rate is doubled when the packet size is halved.

Third, the average latency of the four DDR4 measurements is rising with the packet size. This is plausible as the FPGA has a *store-and-forward* behavior, which means the packet sending is not started before the last bit of the packet is received.

5.2.3.5 Subscriber Interference

One main reason for per-subscriber traffic shaping in Internet access networks is the isolation of flows. This means the data traffic of subscriber A should be isolated and independent of subscriber B. In Fig. 5.23, the evaluation setup is shown, considering two subscriber flows in the residential Broadband Network Gateway (BNG) scenario.

This evaluation scenario consists of two flows: (1) Background traffic that is shaped in the QoS co-processor to a constant rate, e.g., 9 Gbit/s. (2) A low rate probe packet flow, consisting of small 178 byte packets with a rate of 2,000 *packets/s*. The purpose of the first flow is to disturb the queueing system. The P4STA framework captures only the behavior of the second flow. The expectation of this experiment is that in the case of perfect subscriber isolation, no change in the behavior can be observed depending on the background traffic.

The results of this experiment are shown in Table 5.10. First, we measured a baseline with zero background traffic, with an average latency of 7.24 µs. The average latency increases slightly for a background load of 1 and 9 Gbit/s, as visualized in Fig. 5.24. For 9 Gbit/s background load, latency peaks up to 14.47 µs are present.

However, the increase in average and maximum latency is very low and can be neglected compared to the current end-to-end latency in the Internet, far above 1ms (compare Sect. A.3). The utilized FPGA design for the QoS co-processor consists of a single DDR4 memory and no internal SRAM-based memory for packet storing. Therefore, the two flows share the same memory interface, which can cause

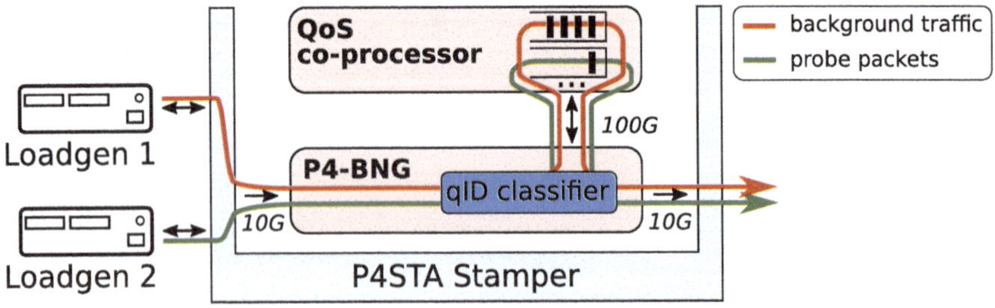

Fig. 5.23 Evaluation scenario for subscriber isolation in residential access networks

Table 5.10 Background traffic influence on the P4-BNG + FPGA system. 178 byte probe packets, e.g., VoIP packets, are sent with a constant rate of 2,000 packets/s in the test queue. In another queue, a constant packet stream is shaped at 0/1/9 Gbit/s

Background traffic:	# packets	Loss	**latency**			
			Min. (μs)	Max. (μs)	Average (μs)	Std. dev. (μs)
None	10,000	0	2.02	12.85	7.24	3.00
1 Gbit/s	10,000	0	2.01	12.76	7.39	3.00
9 Gbit/s	10,000	0	2.01	14.47	7.34	3.19

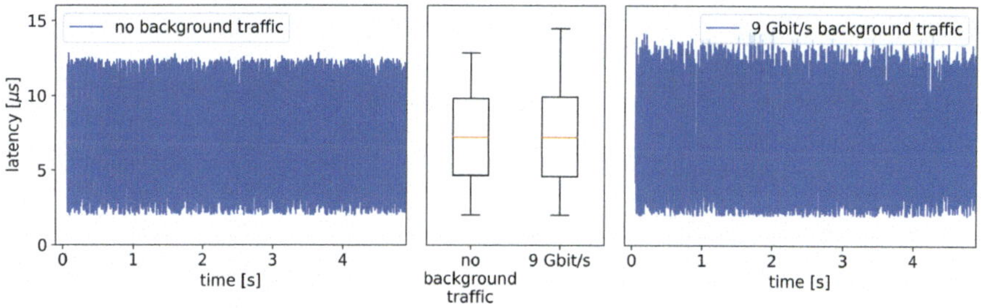

Fig. 5.24 Influence of background traffic on the FPGA-based traffic shaping for no background traffic and 9 Gbit/s background traffic, integrated in a residential BNG system. Remaining measurement setup as in Table 5.10

collisions and micro congestion while writing and reading data. We assume that this shared memory interface is the cause of the slight latency increase of $\leq$150 ns.

Overall, we can conclude that almost no subscriber interference occur in the presented QoS co-processor design. Therefore, the goal of subscriber isolation at the Internet access edge is achieved.

5.2.3.6 Resource Utilization

In this thesis, we discussed multiple configurations of the FPGA-based QoS co-processor. In the following, the utilization of the resources on the FPGA is presented. Note that the presented results are generated only for Xilinx-based FPGAs; however, the general trends are universal for FPGA-based QoS systems relying on other vendors' chips.

FPGAs consist of multiple resources, which are all utilized to realize the described boolean behavior.

Lookup tables (**LUT**) are used to represent the combinatoric logic of the digital signals. In a Flip Flop (**FF**) memory, a single bit can be stored. With *LUTs* and *FFs*, the most common boolean logic can be realized, with except of large memories.

Table 5.11 Resource utilization of the QoS co-processor for a Xilinx Alveo U200 FPGA. In all configurations, a single 100 Gbit/s Ethernet IP core for receiving and sending packets is used

Configuration	LUT	FF	BRAM	URAM
2048 queues, 16384 buckets,	36,463	51,104	130	6
1x DDR4, no-scheduler	(3.1%)	(2.3%)	(6.0%)	(0.6%)
2048 queues, 16384 buckets,	36,775	54,445	130	8
1x DDR4, simple-scheduler	(3.1%)	(2.4%)	(6.0%)	(0.8%)
2048 queues, 16384 buckets,	37,178	54,596	130	9
1x DDR4, simple-scheduler, AQM	(3.1%)	(2.4%)	(6.0%)	(0.9%)
2048 queues, 16384 buckets,				
2x DDR4, 1x SRAM,	65,373	88,797	157	73
Simple-scheduler, AQM	(5.5%)	(3.8%)	(7.3%)	(7.6%)
262144 queues, 131072 buckets,				
2x DDR4, 1x SRAM,	66,576	89,108	173	416
Simple-scheduler, AQM	(5.6%)	(3.8%)	(8.0%)	(43.3%)

Within the FPGA, internal Static Random Access Memory (SRAM) blocks are available, named **BRAM**. In addition to *BRAM*, "ultra RAM" (**URAM**) blocks exist. Despite their larger size, they are similar to *BRAM*. In the following, we focus on these four resources. However, many more hardware primitives exist on FPGAs, which are typically not limiting.

The resource utilization for multiple configurations is shown in Table 5.11. They are an output of the FPGA synthesis tool and are specific for the used Xilinx Alveo U200 FPGA. In the first row, the results for a minimal evaluation configuration are shown, consisting of only a single external DDR4 memory and the *No-Scheduler*. The utilized resources are pretty low and already include the peripheral infrastructure, i.e., the IP cores for PCIe, Ethernet, and the external memory.

In the second row, the *No-Scheduler* is exchanged with the *Simple-Scheduler*. We note an increase in the *URAM* utilization (plus two blocks) caused by the additional state memory of the token-bucket for 2048 rate-limited queues. Similarly, in the third row, the added Active Queue Management (AQM) instance causes an additional increase of *URAM* (plus one block).

From this, we can infer that the scheduler and AQM functionality requires additional functionality while the other resources remain almost unchanged. The influence by the scheduler is two times higher than for the AQM algorithm, as the scheduler stores a token bucket value and a rate value for each queue, 54 bits in total. On the other hand, the AQM instance requires only 32 bits per queue, fitting with the data width of a single *URAM*.

The same configuration is used in the fourth row, but width two external DDR4 memories and an internal SRAM memory for packet queueing. The significant resource increase of *LUTs*, *FFs*, and *BRAMs* can be explained by the second IP-core for DDR4 memory access. The internal SRAM-based memory block for packet storing causes a very high increase in *URAM*. Therefore, it should be well-considered if this internal memory is really needed.

In the last row, the number of queues and buckets for packet metadata storing is increased to realistic scaling dimensions in large-scale Internet access networks. Here, we can observe almost no additional increase for *LUTs* and *FFs*. However, the utilization of *BRAM* and *URAM* is increased. Especially the *URAM* utilization is at the limit of the synthesis tool, as electrical signals between the logical components must be routed with strict timing requirements. In the proposed prototype, the clock frequency is 220 MHz, which means that no electrical signal within the FPGA is allowed to have a propagation delay of greater than 4.545 ns. If higher resource utilization is desired, either the clock frequency must be lowered, or the logic of pushing and popping packets in the queues must be slowed down.

From these numbers, and after analyzing the hardware description code, we can deviate the following mathematical correlation, describing the memory demand of the queueing logic:

$$\text{memory demand} \sim 5 \cdot \#\text{queues} + 2 \cdot \#\text{buckets}$$

Here, we assume the *Simple-Scheduler* and AQM support. This formula does not consider the peripheral IP cores that are independent of the number of queues and buckets.

In general, we can summarize that the presented FPGA design for QoS functionality requires mainly memory. FPGAs are available in many different sizes with strongly varying allocatable resources. The presented resource utilization numbers allow an assessment of which chipsets may be suitable for this or similar functionality. Note that besides the FPGA internal resources, an Ethernet connectivity must be provided by the FPGA.

5.2.3.7 Control Plane Performance

In this work, we proposed two concepts for control plane integration of the FPGA-based QoS system: (1) Using a *kernel module*, or (2) directly accessing the FPGA from the user space by *memory mapping*. For evaluation of these two approaches, we read and write 32-bit values from and to the FPGA via PCI express. The results are shown in Table 5.12. Each test is executed 100 times, and the average time for a single operation is presented. In addition, we performed the same read and write test over a gRPC control plane implementation, building upon the *memory mapping* approach.

The results show, that read operations require significant more time than write operations. This is not surprising, as in addition to the command, the read result must be provided to the control software.

For the *kernel module* the average time to read or write a value on the PCI express device is higher than for the *memory mapping* approach. In both cases, a user space application performs the operation, but for the *kernel module* approach a context

Table 5.12 PCIe control plane performance: average time for sequential 32-bit configure write and read requests

	Kernel module (μs)	Memory mapped (μs)	gRPC local (μs)	gRPC remote (μs)
Write	5.37	1.25	270.4	14,822
Std. dev. (write)	0.46	0.22	49.84	5,154
Read	11.11	5.67	271.2	15,010
Std. dev. (read)	2.40	3.74	41.46	6,311

switch is required, when handing over the command and data from the user space. This explains the decreased performance. The widespread prejudice that kernel modules are very fast is not right, if they are very often called by a user space application and only for tiny tasks.

Compared to these results, the average time for operations via the gRPC control plane interface is much higher. If all gRPC components run locally on the same server, the average read and write time is ~270 μs, a significant degradation of performance. If the controller runs on a *remote* server, i.e., ~15 ms RTT away, the time is increased further. A the read and write operations are executed sequentially, the network delay dominates the execution speed. Therefore, we can take away that the controller should be located either very close to the FPGA or a parallelism concept must be introduced, i.e., bulk operations. Note that these results are not specific for the presented FPGA prototype, they are universally valid for every PCIe-based hardware that is controlled over a gRPC interface (or similar protocols).

The general behavior, that a high delay on the control plane path lowers the flow rule installation, was discussed in related work before. For example Nguyen et al. investigated the control plane performance of SDN-capable OpenFlow switches [10]. One of their findings is an inverse-linear correlation between the controller to data plane round-trip time and the speed of flow rule installation.

We can summarize, that the control plane performance is currently lowered by the control protocols on the higher layers, e.g., OpenFlow or gRPC. Both investigated FPGA-access methods, the *kernel module* and *memory mapping*, have a significant better performance than the high-level protocols, running on top of this. However, it is noteworthy that the *memory mapping* approach has a better performance and allows a simpler realization as no kernel module must be compiled. Therefore, we hope for kernel module free drivers of next generation networking hardware, which is currently not state-of-the-art in commercial products.

5.2.4 Energy Consumption of Programmable Hardware

One main advantage of hardware-accelerated network functions over pure software-based approaches is the improved performance. Even though very low latency and

Table 5.13 Measured energy consumption of the *DPDK baseline* packet forwarder. Server: Dell R740 with 1x Intel Xeon 4110 and Intel 82599 NIC

	Idle	10 Gbit/s	Δ
Server	77 W		
Server + DPDK running	97 W	97 W	~0 W
Δ	~20 W		

jitter are not likely achievable with software, at least the same throughput could be achieved by many parallel server instances, sharing the total load. However, scaling up the number of servers increases the energy consumption linearly. As Internet access networks are responsible for ~70% of the total network energy consumption, this prospect is significant [13]. According to Betker et al., the total energy consumption of computer networks scales almost linearly with the total traffic volume [2].

In the following of this section, we evaluate the hardware acceleration technologies in this work briefly regarding their energy consumption. Specifically, we consider (1) a basic user-space packet forwarder, discussed as *DPDK baseline* in this work before, (2) a P4-programmable *Tofino* switch, and (3) the FPGA-based QoS co-processor of this work.

All experiments are performed with the same IP-based power outlet *NETIO 4*, offering a power meter functionality. However, we could not precisely determine the accuracy of this measurement tool, and therefore, the results should be seen primarily as relative numbers to each other.

First, we investigate the behavior of the *DPDK baseline*. Table 5.13 presents the measured energy consumption for the Server (1) in idle mode, (2) with a running DPDK application running, and (3) with 10 Gbit/s throughput. The idle energy consumption includes two fiber-optical transceivers, causing an increase of ~2 W. We see that the power consumption increases by ~20 W, caused by a single CPU core busy-waiting for new packets with 100% core utilization. However, if we send 10 Gbit/s test traffic through the system, the power consumption remains constant or below the measuring inaccuracy. A complex network function that operates on multiple CPU cores has an almost linearly increased power consumption with the number of CPU cores. Similarly, multiple network interfaces can be served on a single server, each requesting one or multiple CPU cores.

Next, we focus on the energy consumption of the *P4-Tofino*, which is presented in Table 5.14. In the first two rows of the table, the switching chip, a so-called *Application Specific Integrated Circuit (ASIC)*, is not yet enabled. We observe a power consumption increase of 9 W for enabling the fans at a constant rate of 30%. This fan speed is constant during the experiments and can not influence the results. Note that a further increase of the fan speed significantly increases the power consumption; we measured up to ~250 W at maximum speed. Therefore, the fans can become non-neglectable power consumers. If we enable the ASIC, the power consumption

Table 5.14 Measured energy consumption of the *P4-Tofino* network switch under different circumstances. P4-Switch: BF6064X-T with 2 x 100 Gbit/s copper cable

	Idle	100 Gbit/s	Δ
P4-Switch, no fan	92 W		
P4-Switch, 30% fan	101 W		
P4-Switch, fan, ASIC enabled	166 W	168 W	~2 W
Δ	~65 W		

Table 5.15 Measured energy consumption of the FPGA-based QoS co-processor (Alveo U200). FPGA configuration: 262, 000 queues, 2x DDR4, SRAM, Simple-Scheduler, AQM. Server: Dell R740 with 1x Intel Xeon 4110 and commodity NICs

	Idle	99.9 Gbit/s	Δ
Server	75 W		
Server + FPGA	98 W	108 W	~10 W
Δ	~23 W		

increases by ~65 W. The power consumption increases by 2 W if we send 100 Gbit/s data through two ports of the switch.

Last, we investigate the energy consumption of the FPGA-based QoS co-processor. The measurement results in Table 5.15 show an increase of 23 W for the entire system if the FPGA is physically installed and the appropriate configuration is loaded. The consumption without FPGA is 2 W less than for the *DPDK baseline* (compare Table 5.14) as two fiber-optic transceivers less are installed. In addition, the packet queueing system causes a power consumption increase of 10 W / 100 Gbit/s.

We conclude that all compared approaches have a constant, throughput independent energy consumption. Further, a load-dependent power consumption exists, i.e., *W/Gbit/s*, which strongly varies between these approaches. Thus, hardware acceleration is more suitable for network functions with high throughput.

5.3 Active Queue Management in Programmable Hardware

In Sect. 3.3, we introduced concepts to realize Active Queue Management (AQM) in programmable data planes. More specifically, we investigated the capabilities of P4-programmable switches and FPGAs. This contribution improves the Internet service creation on these hardware platforms, as AQM can enormously improve the service quality in Internet access networks.

In Sect. 5.3.1, we evaluate the concepts of mapping the CoDel AQM algorithm on P4-programmable hardware. Afterward, the FPGA-based realization of the same algorithm is evaluated in Sect. 5.3.2.

Note that AQM algorithms work in a control loop with the congestion control of the transport layer protocol, e.g., TCP. Therefore, the presented results depend on many factors, including the operating system, link delays, interrupt structures, and slightly vary between experiments.

5.3.1 P4-CoDel

In the following, the P4-CoDel implementation, introduced in Sect. 3.3.2, is evaluated. We implemented CoDel on two P4-capable hardware platforms:

- **Intel Tofino:** The prototype on this platform is realized within the P4-programmable egress pipeline of the switch, located behind the packet queues. Thus, the latency of the current packet can be used as input for the AQM algorithm.
- **Netronome SmartNIC:** In this architecture, the AQM algorithm is located before the packet queues. Therefore, the current queue level of the corresponding queue is used as input. Further, this hardware platform has no conventional pipeline. Instead, it relies on a many-core processing architecture.

We compare these implementations with the Linux kernel reference implementation. In Fig. 5.25, the latency of one and three starting TCP flows is shown for the first four seconds. The rate is limited to 100 Mbit/s.

For a single TCP flow, in Fig. 5.25a, the time series is similar for the two investigated implementations and the Linux kernel reference implementation. In the beginning, the latency increases rapidly until the CoDel algorithm starts with congestion prevention. From now on, the TCP sending rate is influenced by controlled packet dropping to fall below the *TARGET* delay of 5 ms. This is the expected behavior and coincides with the Linux kernel implementation.

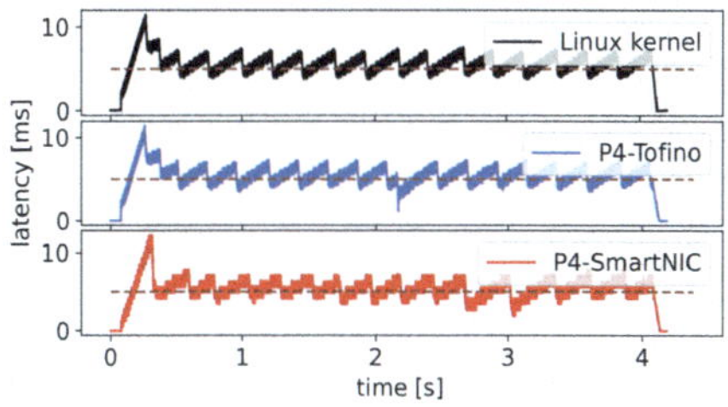

(a) 1 TCP flow.

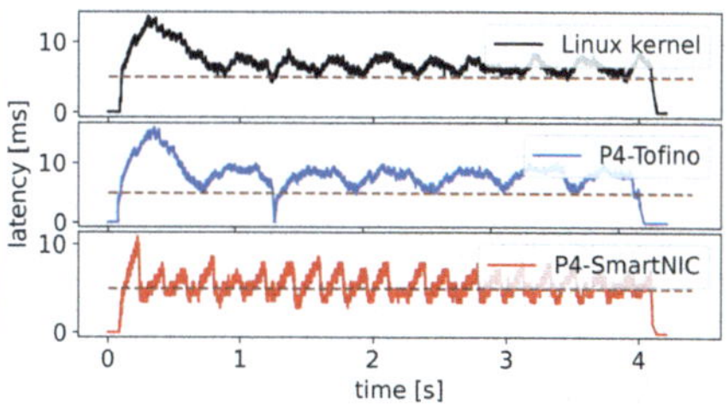

(b) 3 TCP flows.

Fig. 5.25 Latency time series of the CoDel algorithm, realized in two different P4-programmable data planes and the Linux kernel reference implementation

The results for the three TCP flows in Fig. 5.25b are different. First, the Linux kernel implementation has a higher latency increase in the beginning compared to the same implementation with a single TCP flow. This is caused by the three parallel TCP flows, which have a three times higher congestion window increase and, therefore, a faster sending rate increase. Second, the apparent sawtooth behavior of a single TCP flow is not present anymore. The three TCP flows are all limited by packet drops of the CoDel algorithm. Note that no more than a single packet is dropped at a point in time. Therefore, the three flows cannot synchronize with each other, i.e., reducing their congestion window simultaneously. This causes the wavy latency pattern, still falling below the 5 ms *TARGET* periodically.

The behavior of the *P4 Tofino* implementation is similar. In the beginning, we observe a high latency increase and, from there on, the wavy latency. Thus, the algorithm on the Tofino platform works as expected.

The results of the *P4-SmartNIC* deviate from the expected behavior. The pattern is similar to a single TCP flow except for the faster-increasing latency. This arouses the suspicion of TCP flow synchronization, i.e., many packets are dropped at the same time. As multiple processor cores process the packets in parallel, the CoDel drop decision is likely executed on multiple packets simultaneously. From an algorithmic point of view, this is not intended by CoDel. However, the many-core architecture makes it impossible to avoid this race condition and parallel packet dropping.

Last numeric statistics of this experiment with one, two, and three TCP flows are shown in Table 5.16. The packet loss and latency numbers are different, although still in the same range.

We can conclude that the realization of AQM algorithms in the data plane, e.g., as a supplement to Internet service creation functionality, is possible. While the *P4-Tofino* is well suited, the many-core SmartNIC suffers from synchronization problems caused by its internal architecture.

Table 5.16 Number of dropped packets and average latency for three different investigated CoDel implementations. Each run is 4 s long, and has a configured rate of 100 Mbit/s, corresponding to approximately $33 \cdot 10^3$ packets

#TCP-flows	Linux		P4-SmartNIC		P4-Tofino	
	Loss (%)	Latency (ms)	Loss (%)	Latency (ms)	Loss (%)	Latency (ms)
1	0.05	5.7	0.17	5.51	0.16	5.59
2	0.17	6.5	0.24	6.25	0.36	6.86
3	0.33	7.2	0.45	6.42	0.44	8.17

5.3.2 *FPGA-Based CoDel*

In this section, we evaluate the realization of the FPGA-based CoDel implementation, presented in Sect. 3.3.3. The algorithm is integrated into the FPGA-based QoS co-processor for traffic shaping at the Internet access edge. In the previous section, we evaluated the same algorithm in P4-programmable hardware. In contrast, an FPGA integration offers much more freedom, as no P4 language and hardware constructs made for packet header processing must be diverted.

In Fig. 5.26, the latency for the first four seconds is shown for the *Linux kernel* implementation and the *FPGA* realization. In this experiment, three TCP flows are started in parallel, and the FPGA limits the rate to 100 Mbit/s. The behavior of both traces is very similar: The latency increases rapidly until the CoDel algorithm starts dropping packets. After that, we observe the wavy latency pattern in both scenarios.

However, a close examination shows that the periodicity of the wavy latency is higher on the FPGA. The CoDel algorithm contains an optimization, allowing to drop packets earlier if frequent congestion occurs (compare Line 15 of Listing 3.3). This optimization is not implemented in the FPGA prototype, as this would double the maintained state for each queue. The lack of this optimization presumably causes the behavior difference; however, its effect can be neglected in most real-world deployments as it is almost not visible.

All in all, we can summarize that an AQM algorithm can be deployed in the QoS co-processor. The evaluation results show that the behavior of the realized algorithm is similar to the software reference implementation. Therefore, the same effect as by the software implementation on end-to-end flows in the Internet can be expected. However, the proposed approach works at scale with thousands of packet queues and is deployable at the Internet access edge, i.e., at the termination point of residential and mobile subscribers.

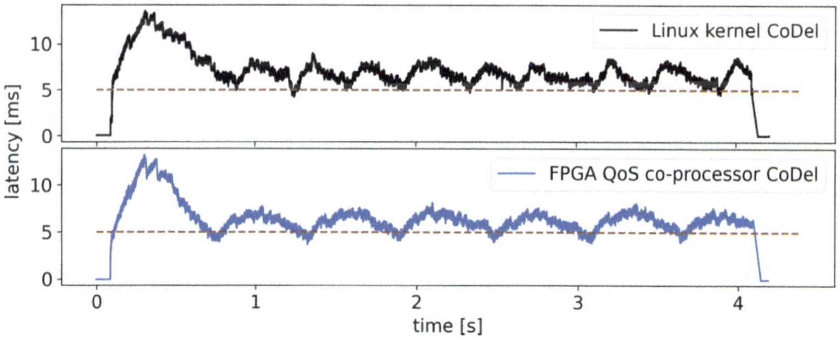

Fig. 5.26 Comparison of the FPGA-based CoDel implementation with the Linux kernel reference implementation for 3 parallel congestion-controlled TCP flows

5.4 Overall Discussion of the Evaluation

In this chapter, we evaluated the contributions according to the first research goal. First, we have shown the performance characteristics of the *host bypassing* approach, enabling a direct data transfer between Network Interface Cards (NICs) and PCIe-based hardware accelerators. In particular, we observed excellent behavior in terms of latency and throughput for our proof-of-concept implementation on Field Programmable Gate Arrays (FPGAs). The Graphics Processing Unit (GPU)-based prototype has shown similar throughput characteristics but was not suitable for latency and jitter-critical applications. However, the *host bypassing* approach has superior performance characteristics for both accelerator technologies compared to state-of-the-art approaches with the same accelerator.

Second, we discussed the performance characteristics of QoS-aware network functions for Internet service creation at the access edge, offloaded on programmable hardware. In detail, the proposed prototypes for residential and mobile Internet service creation on P4-programmable hardware have shown excellent results. We could show the expected behavior of the proposed FPGA-based QoS co-processor up to a data rate of 100 Gbit/s. In conjunction with the P4-based approaches for Internet service creation, this FPGA concept enables a fully functional and programmable hardware basis with the highest performance.

Last, the evaluation of the Active Queue Management (AQM) realizations on programmable hardware has shown the expected behavior, enabling an enormously improved end-to-end service quality in future Internet access networks compared to static-sized queues.

As part of this work and within the evaluation chapter, different programmable hardware platforms, i.e., the P4-programmable Tofino, P4-programmable SmartNICs, Field Programmable Gate Arrays (FPGAs), and Graphics Processing Units (GPUs), were used. These platforms build upon totally different internal structures, benefits, and disadvantages. Here, we would like to highlight the advantages of the programming language P4. While FPGAs can be described with very narrow limitations, e.g., in Verilog, the development process is time-consuming, challenging, and attracts implementation bugs. In contrast, the domain-specific language P4 allows focusing on the functionality in a much faster and easier development process due to the hardware abstraction. Therefore, if the functional expressiveness is sufficient, P4 should be preferred over a low-level design.

In summary, the performance goals of the contributions as part of the first research goal are achieved. The contributions in Chap. 3, aiming for high throughput and low latency realization of network functions at the access edge, have proven their presumptions.

References

1. Accel-PPP, Inc. Accel-PPP: High performance PPTP/L2TP/SSTP/PPPoE/IPoE server for Linux. https://accel-ppp.org/. Accessed 04 April 2022
2. Betker A, Gamrath I, Kosiankowski D, Lange C, Lehmann H, Pfeuffer F, Simon F, Werner A (2014) Comprehensive topology and traffic model of a nationwide telecommunication network. J Opt Commun Netw 6(11):1038–1047
3. Chang C (2016) dpdk_gtp_gateway. https://github.com/edingroot/dpdk_gtp_gateway
4. Đerić N, Varasteh A, Van Bemten A, Mas-Machuca C, Kellerer W (2021) Towards understanding the performance of traffic policing in programmable hardware switches. In: 2021 IEEE 7th international conference on network softwarization (NetSoft), pp 70–78
5. Emmerich P, Gallenmüller S, Raumer D, Wohlfart F, Carle G (2015) MoonGen: a scriptable high-speed packet generator. In: Proceedings of the internet measurement conference. IMC '15. ACM, New York, NY, USA, pp 275–287
6. free5gc (2019) https://github.com/free5gc/free5gc
7. Ibanez S, Brebner G, McKeown N, Zilberman N (2019) The p4-> netfpga workflow for line-rate packet processing. In: Proceedings of the 2019 ACM/SIGDA international symposium on field-programmable gate arrays, pp 1–9
8. Kundel R, Anderweit L, Markussen J, Griwodz C, Abboud O, Becker B, Meuser T (2022) Host bypassing: let your GPU speak ethernet. In: Proceedings of the 8th international conference on network softwarization (NetSoft). Workshop on edge network softwarization (ENS). IEEE, pp 1–6
9. Kundel R, Eryigit K, Markussen J, Griwodz C, Abboud O, Hark R, Steinmetz R (2021) Host bypassing: direct data piping from the network to the hardware accelerator. In: Proceedings of the 14th international symposium on embedded multicore/many-core systems-on-chip (MCSoC). Best Paper Award. IEEE, pp 23–30
10. Nguyen-Ngoc A, Lange S, Geissler S, Zinner T, Tran-Gia P (2018) Estimating the flow rule installation time of SDN switches when facing control plane delay. In: International conference on measurement, modelling and evaluation of computing systems. Springer, pp 113–126
11. Nielsen J. Xilinx Vitis Unified Software Platform. https://www.xilinx.com/products/design-tools/vitis/vitis-platform.html. Accessed 28 Jan 2022
12. Rodriguez D (2018) Next generation central offices transform network edge with datacenter economics, cloud flexibility. Blog post. https://builders.intel.com/blog/next-generation-central-office-transform-network-edge/. Accessed 03 Feb 2022
13. Samdanis K, Rost P, Maeder A, Meo M, Verikoukis C (2015) Green communications: principles, concepts and practice. Wiley, pp 1–419
14. Xilinx, Inc. (2021) UltraScale Architecture-Based FPGAs Memory IP v1.4. Product Guide. https://docs.xilinx.com/v/u/en-US/pg150-ultrascale-memory-ip

Chapter 6
Summary, Conclusions, and Outlook

In this work, we addressed several challenges in accelerating network functions towards high-performance programmability at the Internet access edge. In this chapter, we summarize the content of the previous chapters and highlight our main contributions in the following. Finally, we conclude our work and provide an outlook on potential future works and research fields, building upon this work.

6.1 Summary of the Thesis

In Chap. 1, we motivated this work and described challenges in current computer communication networks, especially the Internet. As the functional and performance demand steadily increases, the underlying networking hardware must grow with these challenges. From this, we derived a need for flexible and programmable network functions with high performance at the access network edge, such as building on programmable hardware. In Chap. 2, we provided background information on current residential and mobile Internet access networks, concepts for network softwarization and programmability, and discussed existing hardware acceleration technologies. Further, we discussed mechanisms for Active Queue Management (AQM) to reduce the end-to-end latency in the Internet, aiming to apply them at the Internet access edge. Upon these analyses and the motivation, we derived three research questions, which we answered in this work:

- **RQ1.1:** How to design, implement, and integrate hardware accelerators, fulfilling the postulated QoS requirements, in networking edge environments?
- **RQ1.2:** Which hardware constraints must be considered for selecting programmable off-the-shelf hardware technologies and algorithms for offloading network functions?
- **RQ2:** How to accomplish the high measurement accuracy requirements with programmable hardware for flexible network function testing?

R. Kundel, *Accelerating Network Functions Using Reconfigurable Hardware*,
Springer Theses, https://doi.org/10.1007/978-3-031-52872-9_6

Based on these research questions, we summarize the contributions and results of this thesis in the following.

6.1.1 Contributions

In Chap. 3, we presented our contributions regarding the first two research questions, *RQ1.1* and *RQ1.2*, aiming to establish QoS-aware network functions on programmable hardware at the Internet access edge.

First, in Sect. 3.1, we focus on the integration of Peripheral Component Interconnect Express (PCIe)-based hardware accelerators in computer systems. This accelerator kind is widely used in mobile Internet access networks. Efficient data input and output to and from the accelerator are essential to achieve high end-to-end service quality. While state-of-the-art approaches suffer from multiple data copies, the presented *host bypassing* approach allows the direct interaction of the hardware accelerator and Network Interface Card (NIC). This concept addresses the research question *RQ1.1*, focusing on the QoS-aware integration of hardware accelerators. The general concept of our approach is to control the NIC directly from the hardware accelerator to prevent unnecessary packet copying. We aimed to improve latency, jitter, and throughput by bypassing the main system memory, built with non-deterministic and shared memory resources. We proposed two different prototypes for Field Programmable Gate Arrays (FPGAs) and Graphics Processing Units (GPUs) with different advantages and disadvantages to investigate the differences between these hardware technologies as part of research question *RQ1.2*.

Second, we investigated how to provide Internet subscriber termination at the access edge with programmable hardware. For this, we presented two concepts for implementing residential and mobile Internet subscriber termination in Sect. 3.2. We investigated three different programmable hardware platforms as part of research question *RQ1.2*, utilizing the domain-specific language P4. In addition, we performed research on the implementation of massive packet queueing systems on FPGAs, aiming to serve thousands of packet queues in parallel. Our novel hardware architecture builds a flexible and generic basis for packet queueing and scheduling in hardware chips. In this context, we investigate the design and implementation prospects of the research question *RQ1.1*. Especially the conjunction of FPGAs and P4-programmable switches turned out to be a very flexibly programmable basis for many network applications, including Internet service creation.

Last, we investigated the practicability of existing AQM algorithms in programmable hardware. These algorithms were initially made for the execution as software of network clients and can lead to a strongly lowered end-to-end latency in the Internet. In Sect. 3.3, we proposed techniques for migrating these existing algorithms to P4-programmable hardware and FPGAs. For this, we applied these techniques to the exemplary algorithm *CoDel*. This contributes to answering the research question *RQ1.1*, how these QoS-aware algorithms must be designed to be mapped on programmable hardware.

Evaluating network functions with very high performance is challenging and requires specialized measurement hardware. Due to the absence of appropriate tools that simultaneously offer all required functionality, mainly a high accuracy and testing flexibility, we proposed the *P4STA* framework in Chap. 4. This framework combined P4-programmable hardware, configured for packet timestamping and loss detection, with commodity software-based load generators. By this functional disaggregation, we combined the performance and time accuracy of hardware and the flexibility of software. Generating test loads in software is crucial for setting up flexible test scenarios. In addition, the software flexibility includes experiment management and results evaluation. By using the P4STA framework proposed in this work, measurement results are automatically documented and easier to analyze.

6.1.2 Conclusions and Results

Upon the investigations related to the *P4STA* framework, we can answer the research question *RQ2*: In order to measure the behavior of high-performance network functions, it is essential to utilize hardware that offers at least the same performance. As it is not cost-effective to build special-purpose hardware at this performance grade, the utilization of universal P4-programmable turned out to be a game-changer. With such programmable hardware, the *P4STA* framework achieved latency measurements with nanosecond granularity at link speeds of up to 100 Gbit/s. This framework served as the primary measurement tool for the contributions of the first two research questions discussed in the following.

In our comprehensive evaluation, presented in Chap. 5, we investigated the presented contributions related to *RQ1.1* and *RQ1.2* regarding their compliance with the postulated performance.

We evaluated the presented *host bypassing* approach in Sect. 5.1 regarding latency, jitter, and throughput. The presented results show that with this approach, significant throughput and jitter improvements and minor latency reductions can be achieved compared to state-of-the-art methods. Bypassing the main system memory, having a non-deterministic performance, and the lower number of memory copies is the reason for this reduction. During our evaluations, we observed a deterministic and low latency up to a throughput of 10 Gbit/s, even at small packet sizes of 300 bytes. Our comparative prototypes for FPGAs and GPUs have shown significant differences: The FPGA prototype has revealed a significantly better latency and jitter behavior caused by the accelerator-internal architecture: While GPUs are mainly a many-core processor architecture and thus have similar advantages and drawbacks as commodity CPUs, FPGAs comprise application-tailored digital circuits with deterministic behavior.

The evaluation of our contributions to Internet service creation on programmable hardware was discussed in Sect. 5.2. Our conceptual studies for residential and mobile Internet service creation have shown an excellent performance. Depending on the chosen programmable hardware platform, we demonstrated failure-free Internet subscriber termination up to 100 Gbit/s. One key contribution of this thesis is the fully functional prototype of a 5G User Plane Function (UPF), which was integrated and tested in real 5G Standalone networks. The behavior of our novel architecture for massive packet queueing on FPGAs has ideal performance characteristics. Zero unexpected packet loss was detected in our tests, and while evaluating the traffic shaping capabilities, no microbursts occurred. We have shown the interference-free isolation of various subscriber flows in multiple queues. To summarize, our evaluation results show that our approach fulfills all requested requirements on the data plane functionality. Thus, programmable hardware, such as P4-programmable switches and FPGAs, are a suitable, performant, and flexible alternative to existing blackbox solutions for Internet access termination. At least one hardware vendor picked up our idea of combining P4-programmable switches and FPGAs and announced a product upon this concept to the best of our knowledge.

Last, we evaluated the implementation of AQM algorithms on programmable hardware in Sect. 5.3. The investigated prototypes have shown the expected behavior, similarly to existing software reference implementations. However, the presented transformation in hardware scales much better for many parallel queues, such as in Internet access networks. Therefore, this approach supplements the beforehand introduced concepts for Internet access creation on programmable hardware, especially to enhance the FPGA-based QoS co-processor. Integrating these concepts in future Internet access edge implementations could enormously lower the end-to-end latency and increase the overall service quality and experience.

Based on the presented evaluation results, we answer the first research question, *RQ1.1*, aiming at the design, implementation, and integration of hardware accelerators while fulfilling the postulated QoS requirements: With the *host bypassing* approach and the FPGA-based QoS co-processor, we presented how hardware accelerators can be integrated much better into the network's data plane compared to the state-of-the-art. Generally, any indirections should be avoided, e.g., copying network packets within a system multiple times without necessity, as the end-to-end performance suffers from the weakest link in the chain. To design and implement network functions at the access edge, fulfilling the favored QoS requirements, the choice of appropriate hardware is mandatory, and the desired behavior is implemented with constantly good performance. For this, we focused on the example use case of residential and mobile Internet subscriber termination. Last, we have shown how AQM algorithms, improving the end-to-end QoS in computer networks strongly, must be adopted to be mapped on programmable hardware. We applied similar concepts to the network function implementation as for the subscriber termination.

The research question *RQ1.2*, aiming at the selection constraints of programmable hardware, can be answered as follows: GPUs turned out to be susceptible to higher jitter and, therefore, they are less suitable for latency-sensitive applications. However, they offer enormous potential for advanced network functions, such as encryption tasks. In contrast to GPUs, P4-programmable hardware has a higher service level in terms of latency, throughput, and jitter. However, the expressiveness of these hardware architectures is rather limited and is suitable only for header processing. We investigated three different specific platforms within the class of P4-programmable data planes, all having pros and cons regarding their performance and functional expressiveness. Last, we investigated FPGAs described on the bit-level for packet queueing systems. FPGAs turned out to be very powerful for this use case but with the drawback of significantly more complicated programming models. To summarize, during the programmable hardware selection process, one should weigh up the required and desired performance needs, functional requirements, and development effort. A single, generally best solution does not exist. However, we can conclude that programmable hardware is a powerful technology class to improve future Internet access edge networks.

6.2 Outlook

The contributions and results of this work contain the potential for further research. First, we did not consider the management of hardware accelerators in virtualized environments at the access edge. It would be beneficial to investigate appropriate concepts to describe hardware accelerators as a resource, managed in a cloud-like manner, e.g., by management tools such as *Kubernetes*.

Second, we considered programmable hardware accelerators to be atomic resources, which means we have exclusive access. However, the shared execution of multiple network functions simultaneously could significantly improve capacity efficiency. For this, existing and novel hardware virtualization concepts must be investigated regarding their applicability to network functions. For example, two network functions operate on two different network ports of a single P4-programmable switch or FPGA, and a third network function can be added at runtime on a third port without interference. Third, we investigated multiple programmable hardware platforms for Internet service creation in this work. However, we noted that they are not exchangeable one by one with ease, and significant modifications of the control plane interfaces were required. Therefore, we see huge potential for research on hardware abstractions concepts to standardize the control plane interaction.

Last, we would like to mention our open-source load generation framework *P4STA*. Even though a commercial product was announced upon our concepts and results, previously presented in a scientific publication by us, this still comprises many open research questions. For example, how programmable network switches, anyway deployed in existing and future data centers, can be used best for benchmarking purposes in operational networks and how to monitor the entire networking.

Open-Source

Scientific work and technological progress are the results of interaction between researchers and engineers. To fortify future achievements of the research community and facilitate the reproduction of our results, we disclosed multiple implementations of this work as open-source projects.

Acknowledgements This work has been supported by the German Research Foundation (DFG) within the Collaborative Research Center (CRC) 1053 - "Multi-Mechanisms Adaptation for the Future Internet (MAKI)." Further, this work has been supported by the Deutsche Telekom AG within the projects "Dynamic Networks 7-10" and by the Federal Ministry of Education and Research (BMBF) within the Software Campus Project "5G-PCI".

Appendix

A.1 Example P4 Program

One essential basis for this work is programmable networking hardware which can be configured by the P4 programming language, as introduced in Sect. 2.3.

In this appendix, we present a simple P4 program to complement the explanations in this work and provide a hands-on example of this language.

The source code in Listing A.1 describes a program that performs packet forwarding based on the Ethernet destination address. A compiler can generate a hardware-dependent configuration file of the P4 switch, FPGA, or other appropriate hardware.

The presented source code describes the behavior of the *bmv2* software switch, a reference implementation to test P4 programs,[1] but the concepts are identical for real hardware switches. This switch consists of four main building blocks, the `Parser1Impl`, `Ingress` pipeline, `Egress` pipeline, and `DeparserImpl`, combined in Line 48 to form a switch. In the preceding lines, these blocks are defined.

First, an ingressing packet at the switch will be processed by the **ParserImpl**, starting in Line 13. The parser is responsible for parsing the incoming byte stream and providing packet header vectors to the subsequent internal pipeline stages. The parser is described as an Finite-State Machine (FSM) and always starts in the state *start*. The *Ethernet* header, including its fields, i defined in Line 1 to 5. This definition is necessary to extract this header in Line 15 of the *start* state. This concept allows describing any existing and future networking protocol. By default, no packet header format is known by P4. Based on the *etherType* of this parsed header, the next parser state is determined (Line 16 to 18), *e.g.*, parsing IPv4 or IPv6 (not shown), or the packet is accepted by the parser, meaning that the ingress pipeline proceeds with this packet.

[1] https://github.com/p4lang/behavioral-model.

R. Kundel, *Accelerating Network Functions Using Reconfigurable Hardware*, Springer Theses, https://doi.org/10.1007/978-3-031-52872-9

Next, the **Ingress** Pipeline describes the packet header processing and is typically implemented in hardware by a staged pipeline with constant packet processing delay. The control flow, i.e., the ordering of tables and actions to be applied, is defined in the *apply* section starting in Line 41. In the presented example, the table *ether_exact* is applied for each packet with a valid *Ethernet* header. This table, specified in Line 29 and below, matches the destination *Ethernet* address of the packet and sets the egress port of the packet by the action *send*. Note that the P4 language is only used to specify the tables; however, they are filled at runtime by a different protocol, e.g., *P4Runtime*, *OpenFlow*, or *gRPC*.

```
1  header ethernet_t {
2    bit<48> dstAddr;
3    bit<48> srcAddr;
4    bit<16> etherType;
5  }
6  ...
7  struct headers {
8    ethernet_t ethernet;
9    ipv4_t ipv4;
10   ...
11 }
12
13 parser ParserImpl(packet_in packet, out headers hdr, inout metadata
       meta, inout standard_metadata_t standard_metadata) {
14   state start {
15     packet.extract(hdr.ethernet);
16     transition select(hdr.ethernet.etherType) {
17       0x0800: parse_ipv4;
18       default: accept;
19     }
20   }
21   state parse_ipv4 {
22     ...
23   }
24 ...
25 }
26
27 control Ingress(inout headers hdr, inout metadata meta, inout
       standard_metadata_t standard_metadata) {
28
29   table ether_exact {
30     key = {
31       hdr.ethernet.dstAddr : exact;
32     }
33     actions = {
34       send;
35     }
36   }
37   action send(bit<9> port) {
38     standard_metadata.egress_spec = port;
39   }
40
41   apply {
42     if( hdr.ethernet.isValid() ){
```

```
            ether_exact.apply();
        }
    }
}
...
V1Switch(ParserImpl(), Ingress(), Egress(), DeparserImpl()) mySwitch;
```

Listing A.1 Exemplary P4 code for the P4 BMV2 software switch

A.2 5G Standalone Laboratory Setup

In this work, we present a novel User Plane Function (UPF) design running upon programmable hardware. In the evaluation of this thesis, we analyzed the behavior of this design in testbed environments with stimulus generators for the relevant interfaces. In addition to this, we set up a 5G testbed to demonstrate the full functionality and to illustrate the end-to-end data transmissions. A working laboratory setup with real User Equipments (UEs), such as commodity smartphones, can guarantee that no major functionality is missing. Concretely, this laboratory setup follows the 3GPP mobile broadband standard in release 15, often referred to as *5G standalone*. In contrast to this, *5G non-standalone* is the mainly deployed 5G technology in current real-world networks. This technology builds mainly upon *4G* technology and protocols that are extended by a second radio channel using the 5G frequencies and modulation techniques. However, the functionality of the 5G core is typically almost unchanged in *5G non-standalone* over *4G networks*.

The main hardware components of this testbed are shown in Fig. A.1 and can be divided into three parts: (1) the UEs, (2) the 5G RAN, and (3) the 5G core. We utilize commodity smartphones from *Huawei*, *OnePlus*, *Oppo*, and *Asus* for the UEs. In addition, two industrial 5G modems were used. The smartphone UEs are connected via USB to a Raspberry Pi microcomputer, which runs a commodity Ubuntu-based Linux. The Android Debug Bridge (adb) allows remote access and screen mirroring these smartphones.

For the 5G RAN, three different realizations exist:

- **O-RAN:** The disaggregated O-RAN consists of one Central Unit (CU), two Distributed Units (DU), and six Radio Units (RU) of the vendor *Airspan*. The fronthaul network allows time synchronization between the RUs and DUs. Further, the *fronthaul switch* ensures low jitter. The DU and CU are deployed on a Kubernetes cluster.
- **Amarisoft Callbox:** The second RAN is manufactured by *Amarisoft* and is a laboratory base station. In addition to the RAN functionality, an IMS server is part of this product and used in this work. It is delivered as a precompiled software, and therefore insights into the internal behavior are not possible.

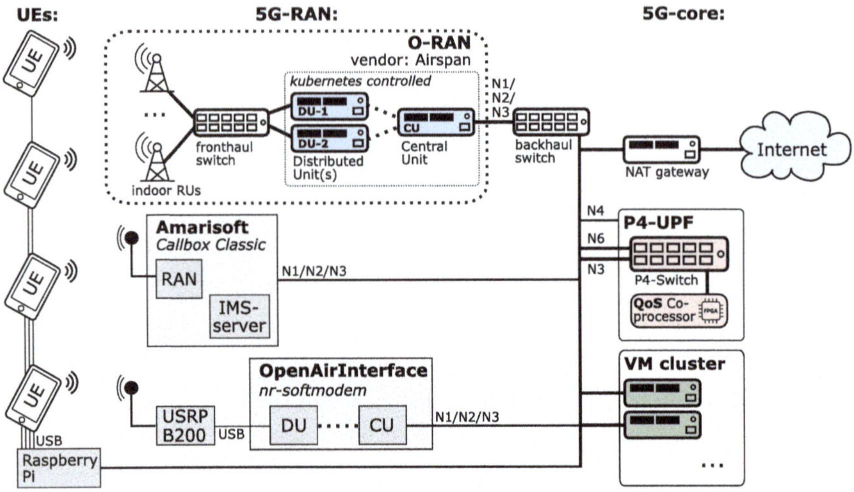

Fig. A.1 Hardware components and simplified network topology of the 5G standalone testbed. Note that not all components are shown to improve the readability

- **OpenAirInterface:** The third RAN is based on the open-source project *OpenAirInterface5G*. A software-defined radio board is used for the radio functionality, attached via USB to a powerful desktop PC. The open-source project can operate in a disaggregated way following the O-RAN terminology with a DU and CU or as a single software component with all required functionality.

All three RAN implementations provide the N1/N2/N3 interfaces towards the 5G core, which the *3GPP* specifies. The *backhaul switch* creates the connectivity, a commodity, VLAN-aware Ethernet switch.

On the right side, the hardware components of the 5G core are shown. The *VM cluster* allows hosting virtual machines for the 5G core and many other infrastructure functions, such as the Kubernetes master for the O-RAN, monitoring functions, a DHCP server, and many more. The *NAT gateway* is a bare-metal server connecting the testbed to the Internet, including subscriber traffic from the 5G UEs. The *P4-UPF* is the design presented in Sect. 3.2.3. Note that either the *P4-UPF* can be used or the default UPF of the 5G core.

A.2.1 5G Core Integration

We chose the open-source *free5gc* project as 5G core, to which we contributed several bug fixes and features during this work [1]. To reproduce the setup described in this section, we recommend version *3.0.6* or later, which contains all required functionality. In Fig. A.2, the detailed setup and configuration of this 5G core are

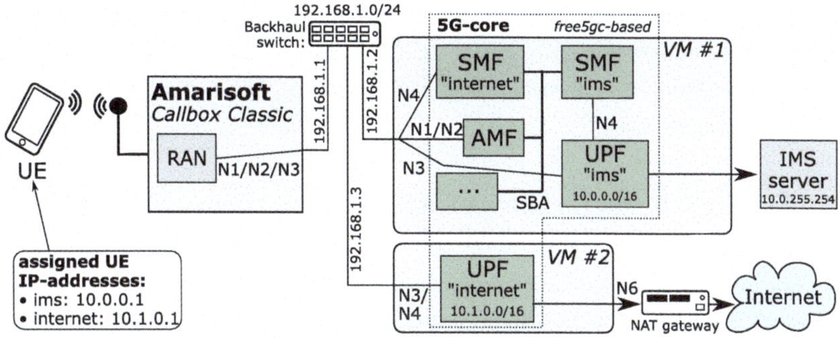

Fig. A.2 Logical structure of the testbed, building upon the open-source free5gc core. Figure derived from: [2]

shown. Note that one advantage of this 5G core implementation is the separated execution of each network function in the core. By this, a single function can be easily exchanged in the Service-Based Architecture (SBA), for example, the UPF. The presented setup consists of two virtual machines (*192.168.1.3*, *192.168.1.4*) and the Amarisoft RAN (*192.168.1.1*). All components are connected to the *backhaul switch*.

In the first virtual machine, all control plane components of the 5G core are running, e.g., the AMF and SMF. Additionally, one software-based UPF is running there for the data network "ims", used for VoIP and described in the following subsection. In the second virtual machine, the software-based UPF for the data network "internet" is running.

If a UE connects to this 5G network, it gets two IP addresses assigned for these two data networks, *10.0.0.1* for *ims* and *10.1.0.1* for *internet* in the example. Packets sent to the data network *internet* are terminated by the second UPF and forwarded to a Network Address Translation (NAT) and from there to the Internet.

The second VM can be replaced by a third-party UPF implementation or by the prototype presented in this work. Only the standardized N4 control plane interface towards the 5G core and the N3 protocol specification must be fulfilled.

Voice over New Radio

The UEs in this testbed setup must support the 5G standalone standard, which is not the case for many tested smartphones regardless of the advertised features set. In addition to various smartphones, we tested two industrial 5G modems successfully. But even for the phones which connected successfully to the network, we observed one main issue: They request a "voice over new radio" service which enables VoIP telephony. If this service is unreachable, they stop the registration process and disconnect. To circumvent this, we added (1) a second data network named "ims", and

(2) started an IMS server. In our tests, we found out that this server must not provide full VoIP functionality; it is sufficient if the UE can open a TCP connection and the server responds with error messages on UE requests.

A.3 Traffic Characteristics of Residential Access Networks

This work aims at the hardware acceleration of network functions at the Internet access edge. Therefore it is beneficial to understand the behavior of Internet access networks, i.e., the traffic characteristics of the packets to be forwarded to and from the subscriber. We performed measurements at a residential access edge of *Deutsche Telekom* in Frankfurt to retrieve these traffic statistics. The measurement was performed over 24 hours in 2019 and captured all subscriber packets in upstream and downstream directions. The exact number of subscribers is not presented for confidentiality reasons but is in the range of multiple thousands and does not influence the key statements. However, in the following, only the results for downstream traffic are shown as they are more meaningful for this work, and most traffic is in the downstream direction.

Figure A.3 depicts the general access network topology of residential Internet access. We captured the packets at the link from the Label Switch Router (LSR) to the Broadband Network Gateway (BNG). Multicast traffic, i.e., IPTV, is not considered as the BNG distributes and duplicates these packets to the subscribers. Therefore a measurement of multicast traffic on this link would not be significant.

A.3.1 Packet Size Distribution

In total, we captured around $54 \cdot 10^9$ packets in the downstream direction. These packets can be assigned in multiple QoS traffic classes, i.e., Best Effort (BE), Low

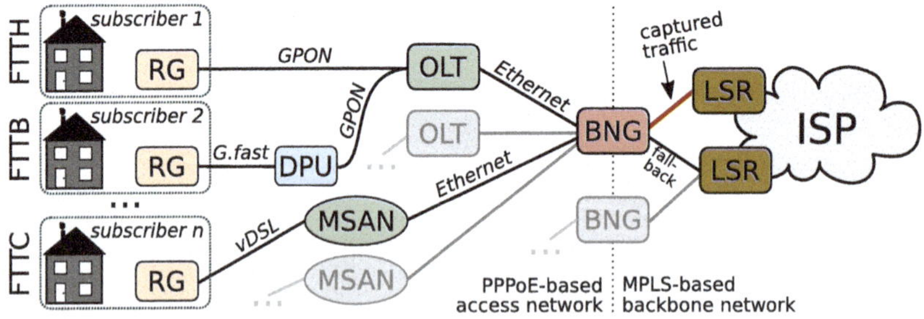

Fig. A.3 Residential Internet access topology, terminated by the BNG. The traffic statistics were captured between BNG and LSR. Figure derived from: [3]

Delay (LD), Low Loss (LL), Voice over IP (VoIP), and network control traffic. Table A.1 presents the distribution of all packets over these QoS classes. We observe that almost all packets belong to the BE traffic class, representing *normal* Internet traffic (Table A.1).

In addition, the average packet size strongly depends on the traffic class, self-inflicted by the applications running in these classes. For example, VoIP traffic transmits the digitized speech in small packet chunks; otherwise, the end-to-end latency would be increased.

Within the BE traffic class, we analyze the packet size distribution in detail, as shown in Fig. A.4 and Table A.2. Here, we classify the traffic into TCP, QUIC, UDP, and other traffic. It is noteworthy that TCP and QUIC are congestion-controlled transport protocols, which means that at least ~93% of the Internet traffic adapts its sending rate if congestion occurs. The packet size distribution has a clear tendency to maximum-sized packets, which means close to the Ethernet Maximum Transmission Unit (MTU) of 1514 bytes. This is no surprise as a TCP data transmission always fills the packets completely, except for the last packet in each transmission. The QUIC transport protocol has a smaller maximum packet size than TCP for unknown reasons (Table A.2).

Table A.1 Packet distribution over the QoS classes

QoS class:	BE	LD/LL	VoIP	Ctrl
Relative share:	99.82%	0.03%	0.14%	0.01%
Avg. pkt. size [byte]:	1314	382	200	886

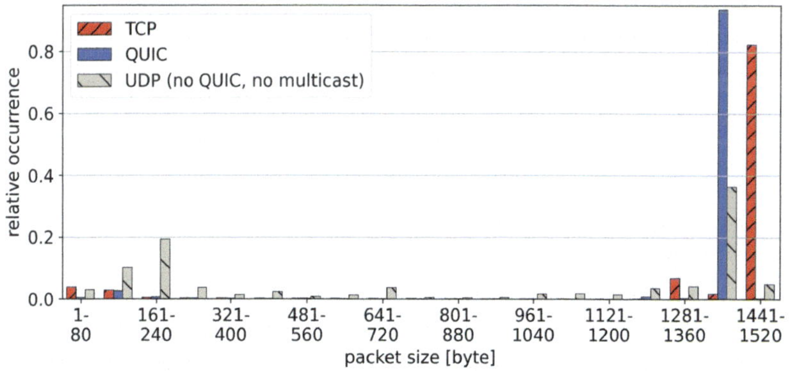

Fig. A.4 Packet size distribution over all observed unicast packets

Table A.2 Unicast packet distribution within the best effort (BE) class

	TCP	QUIC	UDP	Other
Relative share:	84.4%	8.8%	6.4%	0.4%
Avg. pkt. size [byte]:	1347	1319	832	721

A.3.2 TCP Flow Characteristics

The main traffic of current Internet traffic is based on the TCP transport protocol. To understand this more in detail, we captured some TCP flow characteristics. In total, we observed $\sim 70 \cdot 10^6$ TCP flows, each detected by the TCP-characteristic handshake in the beginning.

We can assume no network congestion was caused by this flow during the TCP handshake. By capturing the timestamp of the three handshake packets, we compute the Round-Trip Time (RTT) of this TCP flow. This includes a minor measurement error of the processing time in the TCP endpoints, but we assume this to be neglectable compared to the significantly higher end-to-end latency. In addition to the flow RTT, we capture the flow length F in bytes for each flow, i.e., how many bytes are transmitted from the sender to the receiver until a FIN packet is sent or the flow timeouts.

Figure A.5 depicts the distribution of the flow RTTs in the range from 0 to 112 ms, covering 89.44% of all flows. In addition, the weighted RTT (wRTT) is shown, weighting each flow by its byte length, covering 96.38% of all transferred bytes. These two covering shares indicate that the RTT of flows with a significant number of transferred bytes tends to be lower.

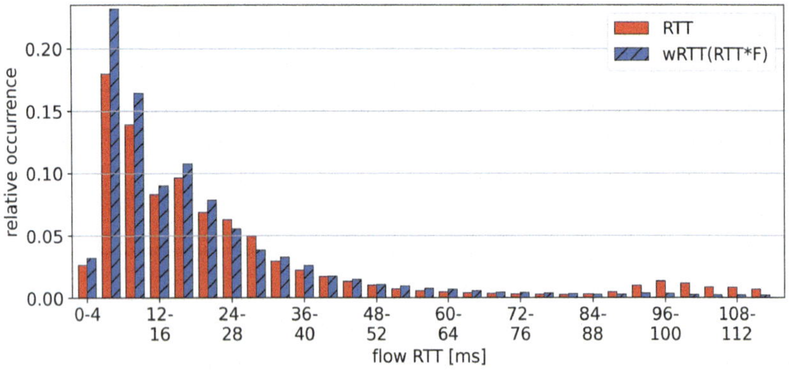

Fig. A.5 TCP flow length distribution of all TCP flows. The weighted RTT (wRTT) weights each flow by its length F in bytes

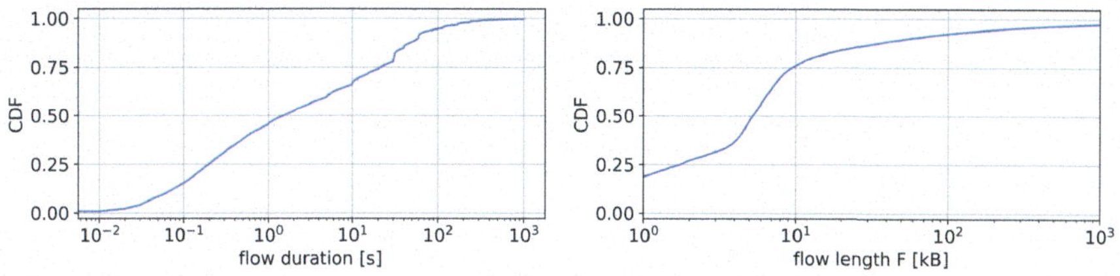

(a) Flow duration until TCP-FIN or timeout.

(b) Flow length distribution.

Fig. A.6 TCP flow statistics for all observed TCP flows

In general, we can postulate that most flow RTTs are quite low, and only a few flows with high RTTs exist. However, most flow RTTs are still distributed between 0 ms and 64 ms. This distribution challenges the static sizing of packet queues, wherefore we discussed the realization of Active Queue Management (AQM) algorithms at the network access edge in Sect. 3.3.

In addition to this, Fig. A.6 depicts the distribution of flow duration and length. We can read from the graphs that 75% of all TCP flows have a duration of $\leq 30\,s$ and $\leq$ 10 kB. The combination of the two plots illustrates that most flows transfer only very little data, or the other way round, a low number of flows transmit most of the data. This is no surprise, as the working principle of most applications in the Internet request many data in parallel in small requests, e.g., domain name requests or loading of html/css/js files. The heavy payload flows are quite seldom compared to these tiny requests, e.g., transferring a large file or streaming a video.

A.4 Token Bucket Accuracy Assessment

In this work, we perform traffic shaping based on the token-bucket mechanism. In the following, we discuss (1) this mechanism, (2) its parameters, and (3) its accuracy depending on these parameters.

The token bucket mechanism works as follows:

- Every T time units, e.g., 1024 clock cycles, new tokens are added to the bucket.
- For each bucket instance i, the number of newly added tokens is $rate[i]$.
- The maximum number of tokens in the bucket is M.

The general update mechanism, applied every T time units, looks as follows:

$$bucket_{t+1}[i] = \begin{cases} bucket_t[i] + rate[i], & \text{if } (bucket_t[i] + rate[i]) < M \\ M, & \text{otherwise} \end{cases}$$

We can freely choose the following three parameters: (1) the time **T**, (2) the **rate[i]** of each queue/bucket instance, and (3) the maximum bucket size **M**.

Accuracy of the Token Bucket Mechanism

To achieve a rate configuration granularity of 1 kbit/s, we can determine the parameters as follows:

(1) 1 kbit/s corresponds to a $rate[i] = 1\,\text{token}/T \mathrel{\widehat{=}} 1\,\text{byte}/T$, the smallest rate > 0.

(2) The time T can be computed as follows:

$$T = \frac{rate[i]}{1\,\text{kbit/s}} = \frac{1\,\frac{\text{byte}}{T}}{125\,\text{byte}/s} = 8\,\text{ms}$$

Assuming a chip frequency of 200 MHz, T is $1.6 \cdot 10^6$ clock cycles in this example. A smaller value T would cause a granularity of the shaper greater than 1 kbit/s.

However, larger values of T cause **Microbursts**, as discussed in the following. We assume a configured rate of 250 Mbit/s. In this case, every 8 ms, $rate[i] = 250,000$ tokens are added to the token bucket. If we assume MTU-sized packets, i.e., 1500 byte, 166 or 167 packets are sent every 8 ms as one microburst. Further, to achieve this rate, the maximum bucket size M must be at least 250, 000, as the bucket would otherwise not be able to hold the added tokens of one round. When T is choosen smaller, this microburst effect decreases, but the granularity is also lowered.

Based on this derivation, we can conclude: A token bucket algorithm has either a fine shaper granularity and tends to microbursts for high rates or has no microbursts for higher rates but a lower shaper granularity. A tradeoff needs to be found. If both are required simultaneously in one system, for each token bucket instance i, a dedicated time $T[i]$ is required, and probably even a dedicated $M[i]$ value.

References

1. free5gc (2019) https://github.com/free5gc/free5gc
2. Kundel R, Meuser T, Koppe T, Hark R, Steinmetz R (2022) User plane hardware acceleration in access networks: experiences in offloading network functions in real 5G deployments. In: Proceedings of the 55th Hawaii international conference on system sciences. ScholarSpace, pp 1–10
3. Kundel R, Wallerich J, Maas W, Nobach L, Koldehofe B, Steinmetz R (2019) Queueing at the Telco service edge: requirements, challenges and opportunities. In: Proceedings of the 1st workshop on buffer sizing. Stanford University, pp 1–6

About the Author

Dr.-Ing. Ralf Kundel was born in Langen, Germany, in 1992. He obtained his B.Sc. and M.Sc. degrees in Information System Technologies from the Technical University of Darmstadt, Germany, in 2014 and 2017 respectively. Further, he received his B.Sc. degree from the same university in business informatics in 2017.

From 2018 to 2022, he was a Ph.D. student at the Multimedia Communications Lab (KOM) at TU Darmstadt, researching the capabilities of programmable hardware for advancing computer networks Access Edges under the guidance of Prof. Dr.-Ing. Dr. h.c. Ralf Steinmetz. He collaborated intensively with Deutsche Telekom during his research, aiming for a disaggregated and powerful Internet service edge leveraging programmable hardware for next-generation networks.

After successfully finishing his Ph.D., he continued working in this field of research in academia and industry. For this, he co-founded the startup kumeco in 2022, aiming to transfer scientific knowledge about softwarized- and hardware-based network technologies from academia to industry.

R. Kundel, *Accelerating Network Functions Using Reconfigurable Hardware*, Springer Theses, https://doi.org/10.1007/978-3-031-52872-9

SPRINGER NATURE

GPSR Compliance

The European Union's (EU) General Product Safety Regulation (GPSR) is a set of rules that requires consumer products to be safe and our obligations to ensure this.

If you have any concerns about our products, you can contact us on ProductSafety@springernature.com

In case Publisher is established outside the EU, the EU authorized representative is:

Springer Nature Customer Service Center GmbH
Europaplatz 3
69115 Heidelberg, Germany

The manufacturer's authorised representative in the EU is Springer Nature Customer Service Centre GmbH, Europaplatz 3, 69115 Heidelberg, Germany. If you have any concerns regarding our products, please contact ProductSafety@springernature.com

Printed and bound by CPI Group (UK) Ltd, Croydon, CR0 4YY
01/10/2026
02225945-0016